A Tribute to the Iconoclastic Mathematician Dr Francisco Bulnes

A Tribute to the Iconoclastic Mathematician Dr Francisco Bulnes

By

Helen Thrift, Michelle A. Musk, Yuri Stropovsvky, Sara Rains, David Cote and Daniel Álvarez-Cruz

Cambridge Scholars Publishing

A Tribute to the Iconoclastic Mathematician Dr Francisco Bulnes

By Helen Thrift, Michelle A. Musk, Yuri Stropovsvky, Sara Rains,
David Cote and Daniel Álvarez-Cruz

This book first published 2023

Cambridge Scholars Publishing

Lady Stephenson Library, Newcastle upon Tyne, NE6 2PA, UK

British Library Cataloguing in Publication Data
A catalogue record for this book is available from the British Library

ISBN (10): 1-5275-5180-6
ISBN (13): 978-1-5275-5180-0

CONTENTS

Acknowledgements viii

Preface ix

Chapter 1 1
Introduction to Dr Francisco Bulnes's Thoughts
1.1. First publications of advanced works, original ideas, perspectives and activities of the seminar "Theory of Representations of Real Reductive Lie Groups"
1.2. Pioneer and Creator of New Theories and Theorems in Mathematical Sciences
1.3. Unstoppable Creative Surges

Chapter 2 15
The Social and Scientific Struggles of a Universal Mathematician
2.1. Living under economic crises in Mexico
2.2. Trips to Tabasco, Campeche, Cotzacoalcos Veracruz, and Tamaulipas
2.3. His friendship with some Russian and Ukrainian mathematicians
2.4. Some achievements and awards around his struggles with education
2.5. Travelling the world and his experiences in culture, science and education

Chapter 3 33
Dr Francisco Bulnes's Cosmovision
3.1. From the Big Bang until the universe today
3.2. Torsion: Its spectra and interactions between fermions of many generations.
3.3. Advanced Results

Chapter 4 .... **44**
The Government, Science and Education: The Tragic Case of Mexico
4.1. Education as a currency of political interests
4.2. External Interests in Mexican Education
4.3. His trips to Cuba
4.4. The Proposals for Improvements in Education in Mexico
4.5. His meeting with the USA: Important Academic Passport
4.6. His trips to Central Europe: Germany, Poland, Slovakia, Serbia, Czech Republic, Croatia, Monte Negro, Romania, and Lithuania.
4.7. More Impressive Recognitions of His Impressive Academic Trajectory

Chapter 5 .... **82**
Expertise and Authority in Several Mathematics Areas, Theoretical Physics and Nanotechnology: A Universal Mathematician
5.1. Authority and Leadership
5.2. Entering some of his theorems in mathematics in geometry, topology, functional analysis, and global analysis
5.3. International Seminars Directed by Dr Bulnes and the Frontiers of Knowledge
5.4. Many of his experiences were shared with colleagues worldwide.
5.5. God's Language
5.6. A meeting of great mathematicians at the Laudal's Fest

Chapter 6 .... **121**
The Nano Development
6.1. On Nano
6.2. "The Conscience Technology on Matter"
6.3. His Mathematical Theory of Nanotechnology: A Review
6.4. A Quantum Algebra Nano

Chapter 7 .... **128**
Cultural and Scientific Development in the United Kingdom and its Relation with Dr Francisco Bulnes in Research and Education
7.1. Approaches to British culture and idiosyncrasies: Its People
7.2. The Fine Popular Music, Some Friends, and Approaches to Modern Science Developed in the UK
7.3. The Work in the UK is beginning to Bear Fruit
7.4. Relationship with the UK and the implementation of its education system in Latin America.

Chapter 8 164
Dr Francisco Bulnes's Research Theory
8.1. Reflections
8.2. Some Theorems and Results

Chapter 9 171
In His Community

Chapter 10 172
Save Our Earth

Technical Notations 174

Acronyms and Nomenclature 177

References 179

ACKNOWLEDGEMENTS

The authors would like to thank Adam Rummens and Helen Edwards at Cambridge Scholars Publishing in the United Kingdom for their application and professionalism during the editing and publication process and their enthusiasm and integrity during the scheduling and establishing of the pertinent strategies for the publication of this work.

PREFACE

This work is a biography of Dr Francisco Bulnes, a world-renowned academician and famous mathematician, especially in Eastern Europe, the Middle East and Asia. This book is written by some alumni and colleagues of Dr Bulnes, who analyse some of his works and contributions. His case is interesting to analyse because he is a very special mathematician, one of the few universal mathematicians, and by universal, we mean that he knows all areas of mathematics (he has an encyclopaedic knowledge of mathematics). Not only that, he has made important contributions to nanotechnology, electronic and electrical engineering, medicine and education. He is like one of those thinkers and scientists dedicated to science and the study of many areas of nature, such as Newton, who was also interested in mathematics, physics and the philosophy of thought and theology.
Francisco Bulnes has stood out as a Mexican scientist, despite the many problems hindering the real and profound recognition of science in Mexican society, more occupied by money, the news of the Jet-Set and world politics than science issues.

The book discusses and explains, first, his work in the fields of mathematics and physics; second, some important facts about his life and their implications for the world; third, his academic friendship with the United Kingdom and what this means; fourth, it contains an analysis of his theorems and natural laws, fifth, the academic awards he has received, and finally, his contributions to education. The book presents his philosophy within mathematics and outside of it, which we analyse and reflect on because we firmly believe that many of his contributions will be much more highly valued and essential twenty or fifty years from now. The world has to evolve and understand much of the academic work of an iconoclastic mathematician like Dr Francisco Bulnes.

Dr Helen Thrift, Author
Research Department in Academic Relations, Cambridge University, UK
Dr A. Michelle Musk, Author
Administrator, UK
Dr Yuri Stropovsvky, Author, UK
Daniel Alvarez, Msc, Author, MX
David Cote, Ind. Eng, Author, MX

CHAPTER 1

INTRODUCTION TO DR FRANCISCO BULNES'S THOUGHTS

1.1. The first publications of advanced works, original ideas, perspectives and activities of the seminar "Theory of Representations of Real Reductive Lie Groups"

Francisco Bulnes, from a very young age, began to show independence in his thinking, his behaviour, his search for knowledge, and his preferences in terms of culture, friendships and study, where in particular, he had a taste for mathematics, which showed him a true and deep knowledge of all things and the entire universe. The choice of mathematics for his career and lifelong pursuit was fueled by these thoughts, which further strengthened as his studies progressed to his PhD in mathematics—a culmination of several of his ideas, having studied the infinite Lie theory, functional analysis and questions of homological algebra and dimension theory. Likewise, he chose and developed his doctoral thesis topic guided by his doctorate director, the great Mexican mathematician Dr Félix Recillas-Juárez[1] [1], with whom he also became close friends. His doctoral thesis is entitled *Some Relations between the Cohomological Induction of Vogan-Zuckerman and Langlands Classification*[2], with the AMS subject in topological and Lie groups [2].

---

[1] A great educator of mathematicians in Mexico (Javier Gomez-Mont, Alberto Verjovsky, José Seade, Enrique Ramirez de Arellano, among others). He was a faculty of science director of UNAM (the National Autonomous University of Mexico). A great international researcher in France, US, and Germany. Some his friends were for example, Henri Cartan, Claude Chevalley (his PhD director), Witold Hurewicz, Hermann Weyl (who taught Félix Recillas in IAS-Princeton), André Weil, Shiing-Shen Chern, Armand Borel, Lê Dũng Tráng, and many more.

[2] At the Institute of Mathematics UNAM, the PhD thesis can be found under the title in Spanish, "Algunas Relaciones entre la Inducción Cohomológicas de Vogan-Zuckerman y la Clasificación de Langlands."

Figure 1. Compilation of lectures and talks given by Francisco Bulnes in IMUNAM from 2000 until 2005. This book was published by IPN, UNAM, ESIME-IPN, SEP and Yucatán University in 2005 during the Appliedmath event.

During his doctoral studies and as part of the doctorate requirements, Francisco Bulnes gave more than 100 talks and lectures on the topics studied in his doctorate, which were presented at the university's Institute of Mathematics. Various volumes of compilations of his talks and lectures were published by IM-UNAM, ESFM-IPN, ESIME-IPN, and CINVESTAV, among others [2, 3-7]. See Figures 1 and 2.

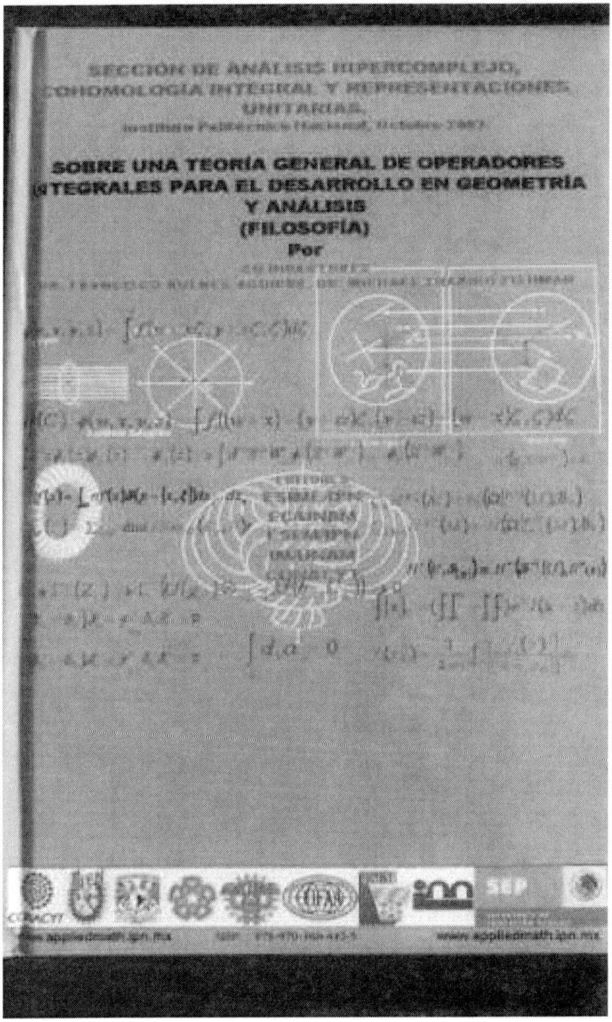

Figure 2. Dr Francisco Bulnes is a co-principal adviser with Dr Michael Shapiro. Dr Bulnes has led several international research groups in mathematics and physics [7-9].

Later, Francisco Bulnes researched several problems of mathematical physics, with a special interest in electromagnetic theory and the analysis of signals and systems; this attracted him much more than functional analysis and the applications in Hilbert and Banach spaces, which, however, did give rise to interesting generalisations and inductions around the study of quantum mechanics, dynamical systems, and telecommunications theory, which he tackled with a systematic study in Euclidean spaces of dimension n, of Fourier analysis.

It is worth mentioning that while he was doing his doctorate in mathematics, he made a treatise from his perspective as a mathematician on the analysis of systems and signals, which was published in 1998 by UNAM, with a prologue written by the great Mexican mathematician Dr Félix Recillas Juárez. In this treatise on superior mathematics, Dr Recillas wrote in the prologue,

> Francisco Bulnes makes a clear and calm exposition of Fourier analysis, considering detailed aspects of mathematical analysis and functional analysis, inviting the reader to a very pleasant journey on these topics with applications to telecommunications, control servo systems and theoretical physics: quantum mechanics and relativity [10]

Francisco Bulnes gave a plenary conference at the Technological University of Mexico in 1994 entitled "Controllability and Signaling", where he presented many of his ideas for the book four years before it was created.

1.2. A Pioneer and Creator of New Theories and Theorems in Mathematical Sciences

Likewise, these preliminary works and others created the scientific and philosophical foundations within mathematics to try and pioneer some new approaches and theories on aspects of studying, from curvature, the most important geometrical invariant of the shape of the spaces and bodies, the Lie groups and their representations, to aspects of study in global analysis, micro-local analysis, deformation theory and derived categories in the study of moduli problems, geometrical ramification problems and geometrical Langland's problems [11-13].

Teaching mathematics courses for engineering (systems engineering, mechanical engineering and electronic engineering) for several years[3], he identified some parameters and invariant aspects of engineering as the final product of mathematics. He began to build and create his mathematical theory of engineering (or TFI[4], directly from the acronym in Spanish (see acronyms and nomenclatures at the end of the book)), in which he sought to standardise and unify the study of engineering from the most important postulate of his theory, that engineering is the creation of energy technology. Likewise, Francisco Bulnes built his first theorems by trying functional analysis tools and applying Lie groups to define finite actions.

However, in the absence of a uniform methodology and the procedural inconsistencies found in some so-called inductive sciences, which can waste time, the impossibility of a specific application in modern science to the traditional methods (such as Bacon's Method), and the lack of effectiveness and efficiency in those traditional scientific methods, made Francisco Bulnes think of extending the mathematical theory of engineering to a systematic theory of research. This was called the mathematical theory of research and was much more extensive since engineering is a specific area of science [14, 15]. Regarding the concept of curvature, Francisco Bulnes continued to carry out further studies, trying to develop the idea of the universe modelled as a complex Riemannian manifold with singularities, where the treatment of these singularities would be through algebraic-geometric methods considering homogeneous line bundles, with the singularities as zeros on the corresponding polynomials of these line bundles [16, 17]. Likewise, complex projective geometries were generated to model the singularities. Through another function, transform solution classes were obtained, which comprised fields corresponding to those sources or sinks in the universe (singularities of the complex Riemannian manifold that models it). But not only that, Francisco Bulnes considered that he could create a good theory of integrals that could build, through the implicit inverse problems and integral geometry, a theory of the universe based on integral classes giving solutions to universal microscopic and macroscopic phenomena. He wrote a book on the general theory of integral operators for the development of geometry and analysis, contemplating from the orbital integrals on Lie groups, the realisations of many representations of topological groups, including the Feynman integrals to describe the field interaction between particles [18].

---

[3] Invited to Cuba to for a postdoctoral stay (2006–2007) after a visit as a council academic delegate in the UCI (Informatics Sciences University of Havana, Cuba) to submit a thesis on applied informatics on the theory of algorithms.

[4] Formal Engineering Theory.

Several years later, he took up research on curvature energy and performed several investigations with the help of electronics and created a new unit of measurement (new metrology), the so-called unit of curvature energy, calibrated on an integral operator called a spherizator (whose nucleus is the Gaussian curvature) and a sensor of curvature fed voltage. Curvature energy is a new concept developed by Dr Francisco Bulnes to control, detect, direct and handle field energy [19-24] with the geometrical invariance inherited in the space that models a physical process, considering its spectra useful in sensing signals in physical phenomena.

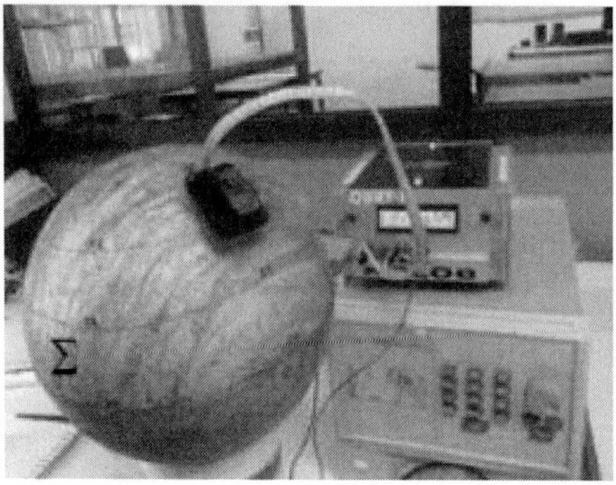

Fig. 3. Gauge sphere used to prove the consistency of the curvature energy unit [23].

Likewise, he created the unit (*Volts/cubic meters*) as a new metrology. This was published in a specialised book on sensors in Europe. It is worth mentioning that Francisco Bulnes has already written and published more than one hundred papers in different international journals on electronics, physics, nanotechnology and mathematics. Also, he has given many international talks at different international events as a speaker, lecturer and plenary lecturer. He has published several book chapters, as well as books and lectures in international research seminars (see Figure 3).

In my particular case, and also of Dr Couper and Dr Michelle Musk, we were interested in the mathematical theory of research because we considered that mathematics is an art form, along with other foundational sciences, as mentioned by Dr Bulnes at different worldwide conferences—in particular the presentation that he gave during one of his many visits to the United Kingdom, where I met him. We began an academic and collaborative friendship.

Applying the mathematical qualities and their essence to research in any area establishes a special paradigm of attention and philosophical innovation of science, where hard-to-explore aspects[5] are due to the impossibility or difficulty of access. Not only that, but the way to obtain substantial advances in studying, being consistent with research, and giving coherence to several ideas as if they were part of a research program, involves much time and effort. Likewise, the most important element of this study is the theory denoted as Th, which must give a foundation to the specific and general investigative tasks and proposes new perspectives and prospects around the object of study.

Thus Dr Bulnes, based on Hilbert's[6] premise that the best application of knowledge is a good theory, extends this premise to the task of research, establishing that "the creation of a good theory is the obtaining of many results, paradigms, applications and conjectures (that is, research for a long time)". Or, as Max Planck[7] would say, the one who makes a lot of effort at the beginning then hardly has to work to obtain new results and paradigms.

Another advantage of the mathematical theory of research with respect to research and technological development is the fact of integrating technological applications as operators t_σ (points of a morphisms group) that are applied on points of a manifold which can be modelled as a Banach or Hilbert manifold to establish a process in technology or technologies. From the point of view of the creation of technologies (this would be the updating of technology), a homomorphism called scientific technologism arises; it is a neologism of the form:

$$\phi(t_\alpha \circ t_\beta) = \phi(t_\alpha) \cdot \phi(t_\beta), \quad (1)$$

where $\circ$, and $\cdot$, are different products in a different space context. One product $\circ$ is for the composition of technologies, and the other, is is for the product of scientific technologisms. One is an operation of the composition of technologies, and the other of the creation process. The law of

[5] Either because the phenomenon is not easy to reproduce, it is in a position that is impossible to access (such as the centre of a reactor), the phenomenon takes place in other dimensions of the universe, or it simply takes place in millionths of a second.

[6] A great mathematician, known as the modern Euclid and the father of functional analysis and integral equations, David Hilbert is considered a giant of the modern mathematics, as big as Gauss or Euler.

[7] Max Karl Ernst Ludwig Planck was a German physicist. He is considered the founder of quantum theory and was awarded the Nobel Prize in Physics in 1918.

composition of the automorphisms α, and $\beta \in G$, takes the form[8] $\xi(\alpha, \beta) = \alpha^{-1}\beta$. Now, the product $\phi(t_\alpha) \cdot \phi(t_\beta)$, is how the product of two energy states makes Dr Bulnes's theory very interesting since producing a new technology should increase the energy level in all its qualities: efficiency, happiness through comfort and ease, execution of the system, and transference ease.

These are all states of energy and are measurable as energy. Therefore, the product of two technology states to produce a third is

$$\phi(t_\alpha \circ t_\beta) = t_\gamma, \quad (2)$$

in the most practical sense, a convolution of technology states. Then scientific technologiscism is, in reality, a neologism of energy, satisfying the principle that engineering is the creation of energy technology. Scientific technologiscism expresses that engineering should be understood as the theory of knowledge for creating technology. Likewise, if Σ, is certain knowledge of a set of true propositions Φ, of a universe of knowledge,

$$\Sigma \subset \Phi, \quad (3)$$

we see that $Th(\Sigma)$ is a theory in engineering if a scientifical technologiscism creates a technology of a certain class. Likewise, let be $\in E = E_{FISMAT}$,[9], $\phi \in \Phi$, and $\sigma \in G$, we have the definition in the engineering context:

$$\phi(\sigma, x) = \phi_\sigma(x) = t_\sigma, \quad (4)$$

Then all technology in any class γ, can be obtained under an engineering homomorphism as:

---

[8] The group G, is a group with structure of Lie group, that is to say, satisfies the group operations of multiplication and inversion, which are smooth mappings (the smoothness of the group multiplication).

$$\xi : G \times G \to G,$$

means that ξ, is a smooth mapping of the product manifold $G \times G$, into G. The two requirements can be combined to the single requirement that the mapping

$$\xi(\alpha, \beta) = \alpha^{-1}\beta,$$

is a smooth mapping of the product manifold into G.

[9] A set or universe of knowledges, mathematics and physics in fact can be considered as our universe, if we consider that this has been designed by an ordered mind.

$$\phi_{\sigma}(t_{\alpha}) = \phi_{\sigma}(\phi_{\alpha}(x)) = t_{\alpha^{-1}\sigma(x)} = t_{\gamma}, \quad (5)$$

Then FET (formal engineering theory) can be constructed, due to Dr Francisco Bulnes. However, Dr Bulnes has gone beyond developing a mathematical engineering theory, considering that applied sciences require engineering to obtain their products. Nevertheless, certain invariants are maintained in the creation of technologies from the applied sciences, and these are the development of a good theory Th, the creation of technology by neologisms ϕ, and the transfer of technology as a homotopic process τ, as defined by Dr Bulnes.

On this last point, technology transfer forces us to question the development of an entity, which can be an individual, a group of individuals, a society, a region, a country, a continent, or even a world in the universe, which can be identified by a knowledge of physics and mathematics.

This is possibly the source of many of the conferences and talks that Dr Francisco Bulnes has given all over the world, where he always questions those attending his talks if mathematics is a creation only of the human intellect or were we designed to understand the beauty of the universe where we live, equipping ourselves with the mathematics to understand it?

This question introduces us to a philosophical problem raised from scholasticism to scientific rationalism and positivism as one of the modern consequences of rationalism.

Dr Bulnes remembers that when he was very young, and especially in the philosophy class taught in high school, the teachers who taught logic and ethics considered him a rationalist due to his inclination towards mathematics. However, Dr Bulnes stated in conversation: "We could be discoverers of the mathematics already inherent in nature, especially in our nature, and we can develop mathematical knowledge by ourselves that helps us develop greater awareness beyond the four-dimensional material universe."

So the higher mind is the garden, and the gardener, too, building our destiny with that higher consciousness acquired through mathematics, metaphysics and quantum physics.

With the development of the formal theory of research and wishing to give consistency to his theory, Dr Bulnes established the so-called consistency theorems, which seek to establish and relate all orders of knowledge (theories, models, technologies, technologisms, prototypes, etc.), including technological products from the central idea of technologism, demonstrating the central problem of the FET. This establishes in depth that models are important for the creation and design of technologies, being, in all cases, the natural models that already exist in the universe, in nature, and that only human beings try to replicate the level of knowledge we have. For example, the aeroplane has been created and designed to emulate birds, although our knowledge about birds is still limited in terms of the efficiency of their flight system, take-off and landing, and the movement of their wings. However, the model remains there, and further developments should continue to consider it as a model to follow. We have the following scheme defined as the second consistency theorem:

$$\begin{array}{ccccc}
\Lambda_{SYSTEM}(\Phi_\alpha(E), t_\delta) & & & & \Phi_\alpha \in Hom(E_{FISMAT}, t_\sigma) \\
 & \phi_\beta \nwarrow & & \swarrow \phi_\delta & \\
\phi_\alpha \downarrow & & \Phi_\beta \in Hom(K_\alpha, t_\delta) & & \downarrow \phi_\gamma \\
 & & & \nwarrow \phi_\sigma & \\
Hom(ThMod\,\Phi_\alpha(E_{FISMAT}), C_n \Phi_\alpha) & \cong & \Phi_\zeta \in Hom(\phi_\sigma(t_\gamma), t_\eta), & & (6)
\end{array}$$

where the last line of the scheme (6) establishes the FET through this isomorphism. Here $Mod,$ is the algebra of models Ψ_j, for some $j,$ $Th,$ the generated theory and C_n, a research unit. Then finally, the research unit of (science and technology) will obtain the technology with a sufficiently solid knowledge base through a set of true propositions and the corresponding models for its design.

Later we will return to this theory of Dr Bulnes. Now let us see his other interesting theories, with their many implications in the deep study of the universe and the projection of future technologies.

Dr Bulnes's energy curvature theory is another way of approaching the field theory problem through its observables and the most important geometric invariant of a space where the field acts. Its energy spectrum is studied, which is constructed through the co-cycles of the cycles of a space manifold and the curvature as a field observable can be detected and measured as an energy signal.

Dr Bulnes uses some integrals developed through his theory to develop integral geometry and obtains new curvature metrology as volts per cubic meter. This was published in a book chapter on sensors [23] and in a research article in the *Journal of Sensor Technology* [21] (see Figure 4).

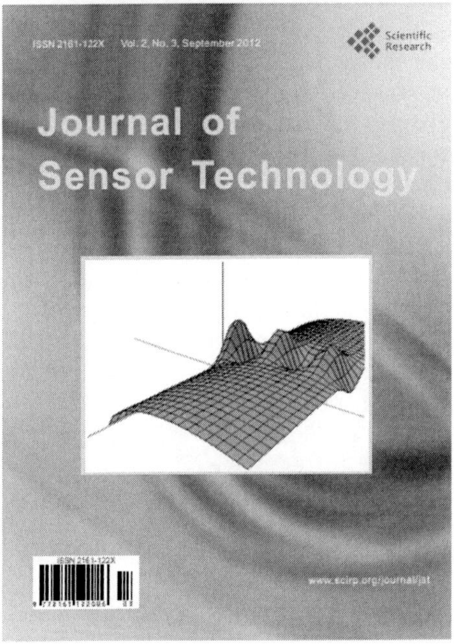

Figure 4. This figure on the front cover is from the article published in the *Journal of Sensor Technology*, 2012, Vol. 2, No. 3, 116–126 by Francisco Bulnes et al.

But not only did Dr Bulnes establish this new concept and theory around it; he also extended it to field torsion, considering torsion as a double curvature. Likewise, he also established that the energy of torsion is its energy of curvature.

Dr Bulnes established different dualities in field theory, geometry and movement to relate the interaction between two fields or movement of bodies in the presence of an electromagnetic field for detecting and measuring torsion. Torsion is a field observable, second curvature in geometry. From the point of view of field theory, torsion is strong evidence of the birth of gravity and its consequences since gravitational waves were detected at astronomical observatories.

Through electronics, he designed an analogue of torsion measurement as evidence of gravitational waves with an experiment to give some insight into what has been studied in the gravitation theories, but with a modern focused study, using invariants as the twistors and spinors for macroscopic and microscopic field theory.

However, the limitations of purely electronic devices only let us see and interpret, using geometrical calculations, certain traces of the electronic signals of torsion, considering a magnetic field is determined in a certain voltage range and a movement of cylindrical trajectory, which, as we know, is constant torsion. However, this verifies conjecture 1.2., and some theorems established in other theoretical and mathematical physics studies. Likewise, a Hall-type sensor detected the magnetic field variation produced from a magnetic dilation [25], which is moved along a cylindrical spiral where the matter agitates the space from which gravity emanates (from the kinematic tensor theory in duality with spinor waves). This also interacts with the magnetic field of the proper dilaton (Figure 5). Francisco Bulnes established and directed this through one of his international research groups.

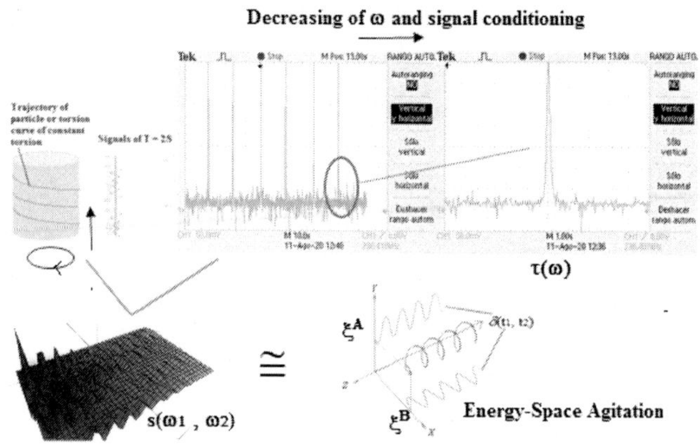

Figure 5. A complete method by the Hall effect sensor for detecting field torsion [26].

The research methods and results are on themes parallel and related to gravity, considering this method is analogous to detecting gravity waves, but in this case, they detect waves of torsion indirectly. If we want to know a complete theory, we have to go back to the theory of curvature on homogeneous spaces that Dr Francisco Bulnes wrote during a semester break while he was the mathematics coordinator of a university. Francisco Bulnes took an agenda given to him and used a draft to write all his theory of curvature on homogeneous spaces as a monograph. Later, a state university, TESCHA[10] and the SEP would publish a book from all these notes [20]. From this book, many research papers in mathematics, electronics, and theoretical physics were published in various specialised scientific journals.

The deep study of curvature in geometry and the development of research on integral geometry by Dr Bulnes, specifically on homogeneous spaces (classes of spaces), led him to derive important technical theorems and lemmas for geometry and the topology of curvature as a tempered distribution. His contact with electronic engineering in the creation of sensors and energy transducers by curvature lead him to establish more conjectures about the nature of torsion and its vital importance in all the processes of the universe. This can be expressed ordinarily as the appearance of waves in space-time agitations. This energy can be an indium of gravity, which scientists measure as gravitational waves when other aspects are involved in field equations [27, 28].

The torsion field mechanism is fundamental for all evolution processes of sidereal objects such as galaxies, black holes, supernova stars, pulsars and nebula. He also observed that this field effect could have relevance and importance for creating future technologies.

1.3. Creativity surges unstoppable

As we have mentioned, the creativity of Francisco Bulnes has been in many areas where he has applied mathematics in addition to his research in pure mathematics. That creativity led him to receive recognition and honour (see Figure 6), not only in mathematics but also in the philosophy of education, by trying to transform education taught at university into a more comprehensive education, where the creativity and imagination of the student are encouraged. For this, it incorporates some specific topics of mathematical research theory, which we have already mentioned, but the aspect of education models with the implementation of research.

[10] Tecnológico de Estudios Superiores Chalco—College of Technological of Higher Studies Chalco, State of Mexico. Here Dr Francisco Bulnes has been a full professor and researcher in electronics engineering since 2006.

Figure 6. Some examples of badges and honours received by Dr Francisco Bulnes, due to his brilliant trajectory. There are more than 50 badges [29-31].

I (Figure 7) will continue with more information and anecdotes about Dr Francisco Bulnes.

Figure 7. Dr Helen Thrift.

CHAPTER 2

THE SOCIAL AND SCIENTIFIC STRUGGLES OF A UNIVERSAL MATHEMATICIAN

2.1. Constantly living under economic crises in Mexico

Francisco Bulnes went through economically difficult times, derived from the recurring economic crises in the country due to poor government in Mexico (he experienced all the crises since 1976 (devaluation), 1982 (the PEMEX failure), 1987 (devaluation), 1994 (the December error), and 2009 (a world crisis), 2019 (the COVID-19 pandemic); and also perhaps due to mismanagement of his money since he used to buy many books and music records when sometimes, he only had just enough left to survive during early adulthood.

While Francisco Bulnes continued his postgraduate studies, struggling with economic ups and downs and working simultaneously while studying for his master's and doctorate, the CONACYT scholarship at that time could not give him enough to maintain a home, a partner or a child.

Back then, the scholarships were not so good, and they had the condition that the scholarship holder must dedicate themselves 100% to study and not to work anywhere. This works for a single person with no children and no commitment, but Francisco Bulnes did not fall into that class, so he had to pay for himself and support his family, his master's degree and his doctorate. However, he received some support for publications of papers, conferences and talks about his doctorate (see Figure 8). He wrote several memoirs of conferences, by IMUNAM, during the doctoral seminar (see Figure 8) [32, 33]. That was a very productive time despite the economic problems and responsibilities acquired.

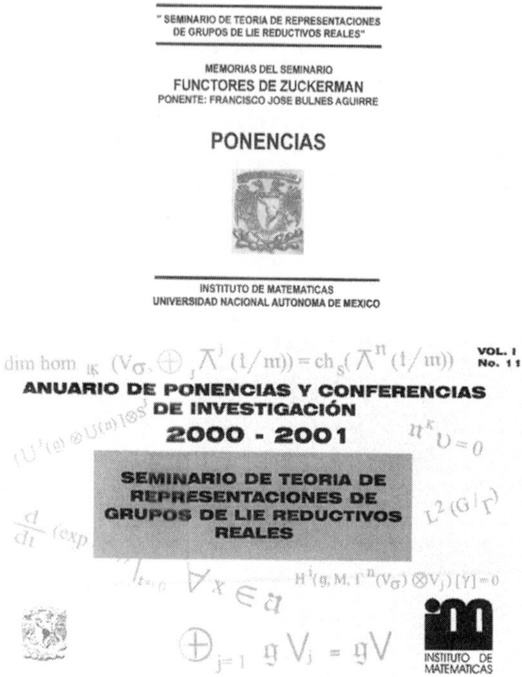

Figure 8. Publications made by the Institute of Mathematics, on presentations and talks given by Francisco Bulnes, as part of the requirements for his doctorate [32, 33].

2.2. The Trips to Tabasco, Campeche, Cotzacoalcos Veracruz, and Tamaulipas

When Francisco Bulnes was Mathematics Coordinator at the Valley of Mexico University (UVM), Lomas Verdes Campus, he travelled the entire east coast of the Mexican Republic, teaching advanced mathematics courses for engineers within the master's program in engineering. systems and computing taught to all staff of PEMEX, AMIME, UTECAM[1], and the Valley of Mexico University (UVM), Villahermosa Campus [34].

---

[1] PEMEX. Mexican Oils. It is a Mexican para-state, which unfortunately represents 80 percent of the national GDP.
AMIME. Mexican Association of Mechanical and Electrical Engineers.
UTECAM. Technological University of Campeche.

Figure 9. Dr Francisco Bulnes (light blue shirt), with some of his master's students from UTECAM (the Technological University of Campeche). Some were officials of UTECAM and others of PEMEX [34, 35].

He toured the territories of Tabasco from Villahermosa to Centla; he also went to Ciudad del Carmen, Campeche, and made a tour of the Zacatal bridge (see Figures 9 and 10) to go to the mainland and attended the Technological University of Campeche (see Figure 9), where he was teaching advanced mathematics to engineers. Some engineers who went there were directors of the oil wells stationed off the coast of Ciudad del Carmen.

During his trips along the Campeche highway, Francisco Bulnes observed the abundant foliage of the jungle that contrasted with the lagoons in different places. Those places were mystical and helped him reflect more deeply on his country. How could the small towns and villages have so much wealth in natural resources and yet be poverty-stricken?

Figure 10. The Zacatal Bridge joins the island of Ciudad del Carmen, which also belongs to Campeche, with the continental part of the state of Campeche itself. This is one of the longest bridges built over the sea and is 3.861km long.

Figure 11. Ciudad del Carmen is a city inhabited mostly by petroleum, civil, electromechanical and computer systems engineers, qualified technicians and oil extraction workers. In the photograph, we can see an oil extraction platform in shallow water. This is about to leave for the open sea near Ciudad del Carmen or the continental part of Campeche [35].

Abundant wealth should directly benefit those populations, and with much of the oil profit should come checks and balances to ensure a certain amount of benefit for the entire Mexican population. Dr Bulnes dreamt of establishing an imaginary corridor from Ciudad del Carmen, passing along the coast of Campeche, Tabasco, passing through Coatzacoalcos Veracruz, and arriving at the port of Matamoros. Good roads, port infrastructure and important airports would benefit the poorer people, the fishermen, and everyone settled in the towns and villages of Campeche, Tabasco, Veracruz and Tamaulipas. Farmers could plant anything on those lands, rich in minerals, and fishermen could sell their fresh snapper and encourage a regional shrimp industry.

Dr Francisco Bulnes met PEMEX authorities and requested a report on the periods of operation, updating of equipment and maintenance of one of the oil extraction platforms stationed in Ciudad del Carmen. He was disappointed as the platforms had equipment from the 1970s, had several damaged pipes, or had fractures that could unleash a fire at any time, and the hydraulic and pneumatic part of the platforms operated poorly, and the automation of some processes was not even carried out. There was an operator for some processes that could have been fully automated. The software could not quickly calculate the compensation required for the dynamic and hydrostatic pressure of the extraction processes, etc. (see Figure 11).

Dr Bulnes proposed research based on the Fourier analysis for implementing a hypercomplex platform for all the automation processes of extraction platforms and other processes derived from or connected with implementing fast calculations on the dynamic pressures. However, as usually happens in Mexico, the proposal was passed on to the corresponding PEMEX directors, at least those from the Campeche, Tabasco and Veracruz areas, without effect. PEMEX's problem, which is the same as always, is how impeded it is by paying civil servants' salaries, excessively high worker benefits, and corruption in its different union associations. There is no intention to invest in modernising and updating PEMEX's engineering infrastructure.

Dr Bulnes established friendships with many platform engineers, mid-level managers and engineers from AMIME (see Figures 12 and 13) and the universities that provided qualified personnel (engineers) to the PEMEX Company.

Figure 12. Here we see Dr Bulnes (in the navy blue shirt) chatting to several engineers doing their master's degrees in systems engineering and computer science. Dr Bulnes taught them advanced mathematics for engineers, including differential equations, integral transformations, special functions, and, in some cases, numerical analysis topics were included.

On the other hand, all the petroleum engineers stationed in Tabasco were also waiting for Dr Bulnes's modernisation project to become a reality. I established a good relationship with them when Dr Francisco Bulnes went to Tabasco to lecture or present a postgraduate course. Many of the engineers wondered if the project would be able to develop since the intention was to tie the professional and specialisation part together in a master's degree in systems engineering, and along with the development of PEMEX, improve its infrastructure and have software and hardware with efficient and effective calculations far superior to that existing at an international level. But the usual thing happened, politics managed everything—that is the tragi-comedy of Mexico.

Figure 13. In Villahermosa Tabasco, with his master's students in systems engineering and computer science, during a meal. Some were officials from Villahermosa's central oil extraction office [34].

Dr Bulnes carried out some very particular projects with his master's students, intending that the seed would remain with them to carry out the proposed systemic modernisation project. There should be better conditions, and the politicians would not tamper with and downgrade technical and scientific work.

Dr Bulnes left a great technical and knowledge legacy in the mathematics applied to engineering among his students from Tabasco, Campeche, Veracruz and Tamaulipas; even a book on his control theory and applications course taught by Dr Bulnes was published by the universities UJAT (Juarez Autonomous University of Tabasco) and UVM-Villahermosa (Mexico Valley University, Campus Villahermosa). Dr Bulnes was invited as a plenary speaker to several UJAT events.

2.3. His Friendship with some Russian and Ukrainian Mathematicians

Obtaining his PhD at such an early age meant that Dr Bulnes had friendships and contact with many mathematicians and academicians from Russia, Ukraine and other Soviet ex-republics, like Georgia, Belarus, Slovenia and Latvia, for many years, principally the mathematicians and scientists from the Russian Academy of Sciences (RAS) and the Ministry of Education and Science of Ukraine (MESU).

The participation of Dr Bulnes in international conferences and specialised mathematics meetings organised by Russia and Ukraine were frequent due to them being free or low cost, and staying in these countries was generally cheap (see Figure 14), not considering the flight costs. But in general, the cost of train travel, meals, and lodging for a scientist from the West was very cheap.

Figure 14. International conferences of topology, geometry and analysis (in Russia and Ukraine), where Dr Bulnes was a speaker and lecturer [36, 37] (also see Figure 16 A).

In addition, the events were of high quality and with the required rigour, being almost always financed by the state government through its ministries of education and institutions in charge of science. Such is the case of RAS (see Figures 14 and 15), which implements the state philosophy adopted since the installation of communism throughout Eastern Europe, guaranteeing free education from nursery age to doctorate level.

Figure 15. Special edition on a mathematics research talk of Dr Bulnes in the Don, Russia [38].

Free education was first established in Russia (in its USSR era), and later, when Soviet rule extended to all of Eastern Europe, it was also adopted by these countries. Today this feature remains a constant. So in times of crisis (where money is scarce), one option for a scientist to continue disseminating their research and continue to get feedback from high-quality scientists is through their participation in these scientific events (Figure 16 A)). Education is a right as important as food, housing and clothing. Francisco Bulnes affirms that this contribution by Russia to the world must be universal.

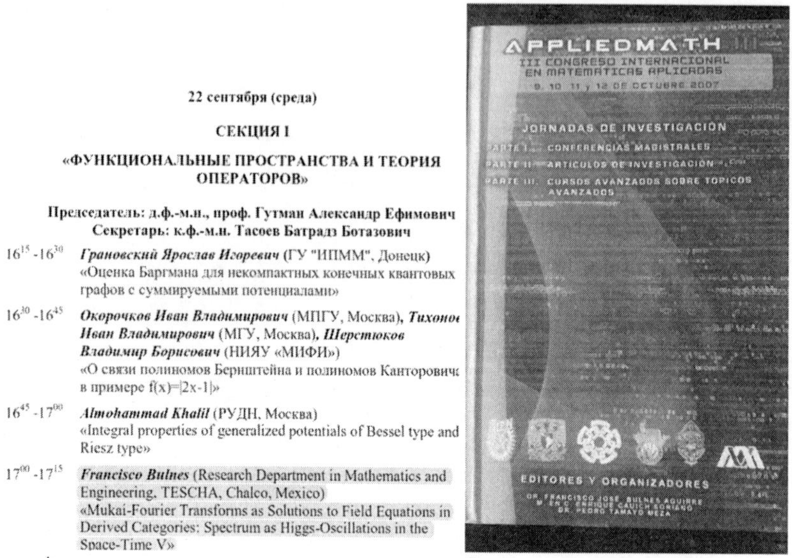

22 сентября (среда)

СЕКЦИЯ I

«ФУНКЦИОНАЛЬНЫЕ ПРОСТРАНСТВА И ТЕОРИЯ ОПЕРАТОРОВ»

Председатель: д.ф.-м.н., проф. Гутман Александр Ефимович
Секретарь: к.ф.-м.н. Тасоев Батрадз Ботазович

16^{15} - 16^{30} ***Грановский Ярослав Игоревич*** (ГУ "ИПММ", Донецк) «Оценка Баргмана для некомпактных конечных квантовых графов с суммируемыми потенциалами»

16^{30} - 16^{45} ***Окорочков Иван Владимирович*** (МПГУ, Москва), ***Тихоно[в] Иван Владимирович*** (МГУ, Москва), ***Шерстюков Владимир Борисович*** (НИЯУ «МИФИ») «О связи полиномов Бернштейна и полиномов Канторович[а] в примере f(x)=|2x-1|»

16^{45} - 17^{00} ***Almohammad Khalil*** (РУДН, Москва) «Integral properties of generalized potentials of Bessel type and Riesz type»

17^{00} - 17^{15} ***Francisco Bulnes*** (Research Department in Mathematics and Engineering, TESCHA, Chalco, Mexico) «Mukai-Fourier Transforms as Solutions to Field Equations in Derived Categories: Spectrum as Higgs-Oscillations in the Space-Time V»

A B

Figure 16. A). Dr Bulnes's participation at an International Conference in Russia [39]. B). International Proceedings of Appliedmath III, directed and organised by Dr Bulnes [40].

Many of the trips that Dr Bulnes made to Eastern Europe allowed him to observe and value some important characteristics of the education model followed in these countries, which was and continues to be the Soviet model of education but without any indoctrination. Dr Bulnes implemented some parts of that model in his research using fundamentals for developing prototypes and experiments.

Dr Bulnes had many friendships with some renowned mathematicians, such as his great friend Dr Michael Shapiro-Fishman, with whom he had some important collaboration for seminars (see Figure 2), and he was an evaluator, judge and dictator of a project called "Hypercomplex Analysis and his Approach to Holomorphic Functions," to be taken to Germany.

Many Russian, Georgian, Slovenian, Ukrainian and Czech academics and mathematicians were summoned and participated in academic events in applied mathematics (Appliedmath I, II, III and IV), organised and directed by Dr Bulnes and sponsored and financed by the Mexican state through the five most important academic entities in the country, UNAM, IPN, CINVESTAV, UAM and Tec de Monterrey (see Figure 16 B).

The fame of Dr Francisco Bulnes as a renowned mathematician, with his high-level research and ability to lead and organise international seminars and conferences, earned him the interest of many PhD students, young, recent doctorates and mathematician colleagues from Russia, Ukraine, Croatia, Romania and even Iran in collaborating on papers for mathematics journals and research advice [41, 42].

2.5. Some achievements and awards around his struggles in education

Francisco Bulnes has made many efforts to make paradigmatic and beneficial educational changes. Some highlights include teaching mathematics in higher mathematics courses for engineers and scientists and the innovative research models he implemented with various international research groups in engineering, nanotechnology, physics and mathematics. One of those methods of his authorship [43-46], developed in multidisciplinary research in several areas, is the method of foundations. This consists of developing an axiomatic-type research methodology, where the guiding axis is mathematics but includes applications for academia and education. The method consists of having an idea or objective for a research program composed of very clearly stated problems; mathematical theories are created from specific theorems, and lemmas arise on the specific research and as consequences of general theorems of mathematical and universal value.

Dr Bulnes received an important international award from India for his professional and postgraduate work as a mathematics professor in the classroom (see Figure 17).

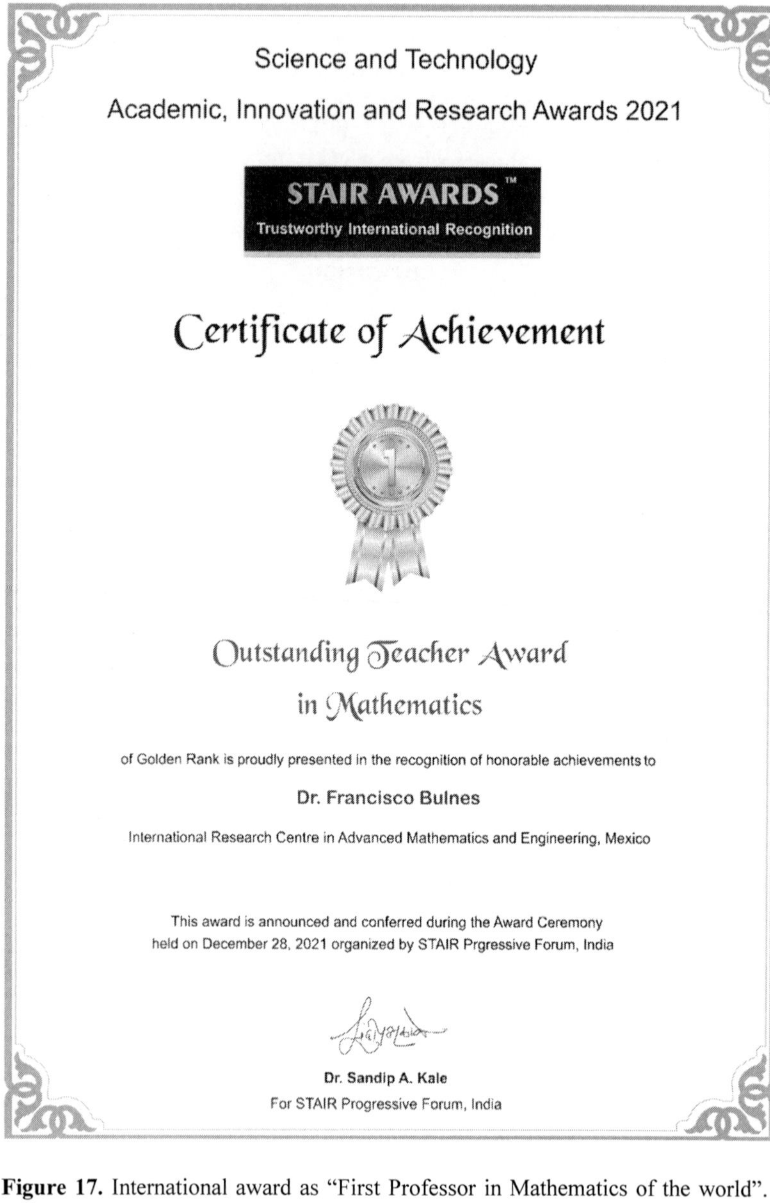

Science and Technology

Academic, Innovation and Research Awards 2021

STAIR AWARDS™

Trustworthy International Recognition

Certificate of Achievement

Outstanding Teacher Award

in Mathematics

of Golden Rank is proudly presented in the recognition of honorable achievements to

Dr. Francisco Bulnes

International Research Centre in Advanced Mathematics and Engineering, Mexico

This award is announced and conferred during the Award Ceremony
held on December 28, 2021 organized by STAIR Prgressive Forum, India

Dr. Sandip A. Kale

For STAIR Progressive Forum, India

Figure 17. International award as "First Professor in Mathematics of the world". This achievement was given by India in 2021 [47].

During his trips to Asia as a delegate from Mexico and director of IINAMEI C.A. (International Advanced Research Center in Mathematics and Engineering C. A.), he had the opportunity to learn about the culture of several Asian countries such as China, Japan, South Korea, Vietnam, Singapore and Malaysia (see Figure 18). He verified that although the countries were all based in Asia, the character and education of each country varies according to their individual long-standing customs, and their traditions continue to differ among them. The Chinese are very similar in character to the Mexicans, although their worldview and traditions are very different.

Figure 18. Dr Bulnes was a Mexican delegate and academician in an international mathematics event organised by China [48].

One of the salvageable ideas in post-revolutionary education in China, that is to say, with the arrival of the leader of the People's Republic of China, Mao Zedong (see Figure 19), was meticulous teaching and the development of education from the elemental degeneration of a concept, creating a "molecular education" in all disciplines, including mathematics, where the latter has had a thousand-year development since ancient China, for example, the Chinese division algorithm or Chinese rest theorem[2]. This

[2] El teorema chino del resto es un resultado sobre congruencias en teoría de números y sus generalizaciones en álgebra abstracta. Fue publicado por primera vez en el siglo III por el matemático chino Sun Tzu. Translation: The Chinese reminder theorem is a result on congruencies in number theory and its generalizations in

peculiarity observed by Dr Bulnes in Chinese education helped him establish his method of foundations, specifically the creation of classes of "inquiry operators" in research theory as part of his mathematical research theory.

Figure 19. Mao Zedong, leader of the People's Republic of China. Propaganda to promote literacy for all Chinese people [49, 50].

According to Dr Bulnes, the leader of Communist China made great contributions to the re-education of the Chinese people, who were immersed in 'ignorance' before the popular revolution promoted by this leader. Mao Zedong understood, like other leaders of communism, that the way to order a country is first to order the ideas and spirit of a country, which must first be educated, starting with the most elementary part, its literacy (see Figure 20).

---

abstract algebra. This was published for first time in the Century III A. J, by the Chinese mathematician Sun Tzu.

Figure 20. Dr Bulnes walks through an ordinary neighbourhood in the suburbs of Shanghai, China. The Chinese people come and go carrying out their daily tasks.

The people are introduced to the arts, science and culture, respecting China's own ancient culture, which has a worldview very different from the Western one and is very interesting. As we have mentioned, the peculiarity of education in China under the scheme of a "molecular method in education" where the formation of the student is stitched together little by little like a beautiful silk rug, with the patience that has characterised them (since Confucius), until the learner is re-discovering and acquiring knowledge is something that Dr Bulnes has rescued to implement in his fundamental research work. This gave Dr Bulnes great fame in Asia, which was transmitted to the Middle East through the common geographic borders of China with Pakistan, India, Uzbekistan, Kyrgyzstan, Bhutan, Myanmar, Vietnam, and Cambodia, also to the north in North Korea and Mongolia. In India, Dr Bulnes is part of an international foundation committee dedicated to the fiduciary support of young researchers (see Figure 21).

Figure 21. Research Foundation of India's 2020 international conference on Women Empowerment & Sustainable Development. An international committee member is Dr Francisco Bulnes, Tescha Dept. of RMEC State of Mexico [51].

Dr Bulnes gave several talks on his work in mathematics at different stages in China and other countries of Asia (see Figure 22).

Figure 22. Dr Francisco Bulnes at an international applied mathematics conference in Shanghai, China, in April 2014. His talk was on the applications of algebraic geometry and mathematical physics in space-time studies [48].

During one of these trips, Dr Bulnes had the opportunity to enjoy the traditional dances of China and see the beautiful dancers and their typical folklore (see Figure 23).

Figure 23. Dance was offered to the guests of the international mathematics meeting.

CHAPTER 3

DR FRANCISCO BULNES'S COSMOVISION

3.1. Between Topological Diffeomorphisms, Embeddings and Tensors: Dr Bulnes's Universal Studies

I am Dr Heather Couper, an astronomer from the United Kingdom. I became aware of the existence and fame of Dr Bulnes when I was invited as an academic peer to evaluate an article on theoretical physics by him. He touched on astrophysics issues and some observational data to confirm his theoretical assertions. The article was entitled "Local diffeomorphisms and smooth embeddings to gravitational field II: spherical symmetry and their breaking in space-time," which was sent without the author's name(s), ascription, or country. The latter was for the evaluation to be as objective as possible and to give my opinion, only focusing on the quality of the work presented.

At first, the article seemed very mathematical to me, and I encountered some things that I did not understand or seemed very technical about mathematical cosmology and even beyond the usual field theory used in astrophysics.

However, it seemed an interesting proposal on how spherical symmetry in gravity could be explained; in addition, Dr Bulnes added other approaches to gravity from a microscopic level and the spectrum of its curvature, something I had not heard of.

It also dealt with homogeneous spaces in mathematics, which I did not know about. However, a colleague from the University of Oxford explained that they were class spaces that represented some type of lateral action. So to evaluate the article, I had to ask others about some concepts of complex differential geometry, differential topology, quantum mechanics, quantum astrophysics, gravitation, general relativity, etc. In the end, I needed the opinion of several colleagues, who only knew one part and ignored the other. It was not easy to evaluate a work where many theories converged. After several weeks, I learned from the scientific journal *Physics & Astronomy International Journal*; that the paper had been published in Volume 2, Issue 1 in 2018 [52].

From then on, I was interested in having some communication with Dr Francisco Bulnes in order to look for some collaboration in the future. I have to say that I am a retired astronomer and that I only dedicated myself to writing books on astronomy and astrophysics for university use[1] with my great friend and astronomical photographer Nigel Henbest, including many photographs of the universe that I took from my telescope and the National Space Centre in Leicester. Therefore, talking to a mathematician interested in cosmology and creating his own universe theory was very refreshing, with the possibility of developing some philosophical discussion about the universe with him.

However, we hardly had any live conversations, just several emails and two Skype meetings, where we discussed the role of space-time singularities as a way to measure the ageing of the universe, also the role of dark energy in the material evolution of the universe, something derived from that article by Dr Bulnes that he had reviewed for the corresponding journal. Francisco Bulnes mentioned working on field torsion and its inherent role in interacting electromagnetic fields with gravity. He showed me that he had been working on field torsion for some time, although he wanted to show the spectral focus of torsion as a second curvature, that is, a curvature further amplified by the interaction not only of gravity in space but also by electromagnetism (see Figure 24).

3.2. Some Results of His International Seminars

He had recently published an article with some Russian academics on spinors, torsion and curvature energy, considering the twistor framework (Figure 24).

I did not understand anything since my study was more empirical and observational through observatories with powerful radio telescopes.

Dr Bulnes invited me to collaborate on a book (see Figure 24 and Figure 25). I apologised, as I had several commitments to publish books on the study of the universe focusing on education and the most common readership, i.e. non-experts. I also had some British television interviews and other astronomy and astrophysics projects, including some with NASA (see Figures 26 and 27). During the short time I communicated with Dr Francisco

[1] Couper wrote dozens of books, mostly with Henbest, including three popular titles for children: *Black Holes* (1996), *Big Bang* (1997), and *Is Anybody Out There?* (1998). Her recent publications include *The Universe Explained: A Cosmic Q&A* (2018) and the 2020 edition of Philip's "Stargazing Month-by-Month Guide to the Night Sky." The pair also wrote a monthly astronomy column for the *Independent* from 1987 until early this year (2023).

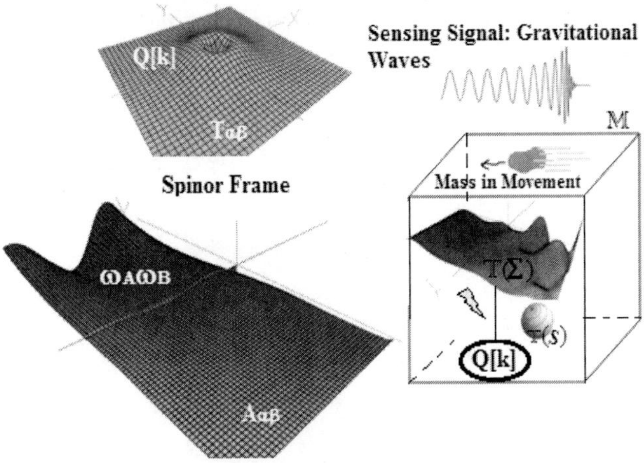

Figure 24. Dr Francisco Bulnes observed the fundamental relationships between twistors, spinors, kinematic tensors, torsion and curvature tensors [53]. These images are part of a chapter of a book edited by Dr Francisco Bulnes: *Advances on Tensor Analysis and their Applications*. This book was published in the United Kingdom in 2020[2].

Bulnes, I believed that his universe studies focused more on cosmology, explaining and modelling the universe and all its phenomena. I believe that he intended to develop a theory of the universe considering a field theory study framework in all its branches, for example, TFT, QFT, QED[3], and those obtained in recent years incorporating string and superstring theory, modern relativity studies, such as the Cartan-Einstein theory, and other

[2] As Dr Couper had died seven months earlier, she cannot have seen this communication.

[3] TFT, Topological Field Theory. This is a quantum field theory which computes topological invariants.

QFT, Quantum Field Theory. This studies aspects of the quantum field.

QED, Quantum Electrodynamics. This is a theory of particle physics, which is the relativistic quantum field theory of electrodynamics. In essence, it describes how light and matter interact and is the first theory where full agreement between quantum mechanics and special relativity is achieved. QED mathematically describes all phenomena involving electrically charged particles interacting by means of an exchange of photons and represents the quantum counterpart of classical electromagnetism giving a complete account of matter and light interaction.

more specialised studies such as BRST-cohomology[4], and referent studies of the universe through MHD. At least it was all these theories that I saw, in addition to other purely mathematical ones, when I evaluated and reviewed his paper.

Figure 25. *Advances on Tensor Analysis and their Applications,* edited by Dr Francisco Bulnes, published in the United Kingdom, 2020 [53].

In summary, I can say that Francisco Bulnes's cosmovision leads us to a universe model that allows us to see the order of things, the development from its birth and evolution to the present day where the topological invariants (see Figure 25) are the fundamental part from which all the descriptions that we can make of the universe derive. This is something that Dr Bulnes loved since, as a good mathematician, the hierarchical order of antecedents and consequents is always defined by the general structures that define and determine the functioning of an entire model.

The rules and forms of operation, action and value of the objects and phenomena that reside there come from these general structures.

---

[4] BRST-Cohomology, in theoretical physics, is a BRST formalism, or BRST quantization (where the BRST refers to the last names of Carlo Becchi, Alain Rouet, Raymond Stora and Igor Tyutin) denotes a relatively rigorous mathematical approach to quantizing a field theory with a gauge symmetry.

Figure 26. The James Webb Telescope reveals extraordinary photographs of nebulas, galaxies, stars and the extragalactic universe. Here is the Carina Nebula [54].

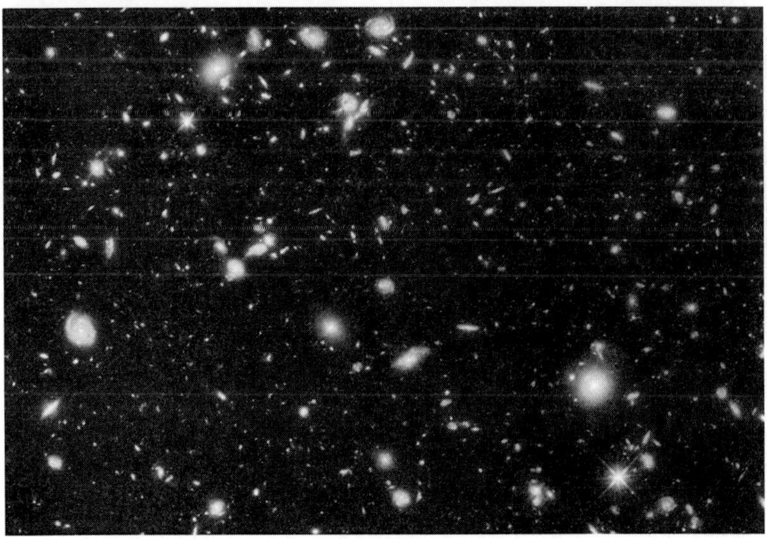

Figure 27. The Hubble Ultra Deep Field images take us back to 400–800 million years after the Big Bang. The extragalactic universe [55].

Another pronouncement of Dr Bulnes, which he has coined and transmitted to his students from his research charts, is that the future technologies of humanity are out there in the cold and inhospitable universe and that trying to emulate the phenomenology of the universe and obtaining the principles of those future and advanced technologies is essential for our survival (see Figure 28). The life systems we know on Earth took thousands of millions of years of physical evolution and development.

From the point of view of Dr Bulnes regarding mathematics, life represents the development of very rich structures defined through symmetric operators in non-symmetric algebras, Hilbert spaces, Hermitian forms, periodic convolution products, spinors, sphere homotopies, Hardy and Bergman Spaces, Sobolev Spaces, and tensor derived categories [56, 57].

Figure 28. Dr Bulnes gave a masterful talk to ESIME-IPN at its Postgraduate Unit 2007.

I am an astronomer (Figure 29), and I find it difficult to say something about the universe if I have not observed it. But I can see that Dr Bulnes does not need to see it. Later I learned that he was focused on field observables and had created a research program on field torsion and its applications. I learned that the United Kingdom published a book written in tribute to Dr Francisco Bulnes, which highlighted his concern for the necessity of understanding the phenomena of the universe, its interstellar structure and particle physics and trying to introduce these processes to future technology. This seems very logical to me as I also believe that all the answers we want to know as humanity about what we are, where we come from and where we are going are in the universe. I also learned of new highly technical publications in astrophysics, cosmology and even some very fine experiments developed with electronics, which Dr Bulnes and his research group had been developing (see Figure 30). Something that surprised me was learning that he was getting some prototype devices that measure and detect curvature and torsion. I asked myself, how can this man lead from such abstract mathematical results and others applied to the formal study of the universe (for example, cosmological theories) to the development of electronic devices to detect field observables, where very fine instrumentation is required to have evidence of the phenomenon studied?

Figure 29. Dr Heather Couper [58, 59].

Figure 30. Physics and Astronomy—2020—Annual Scholars' Hall of Fame, From the United Kingdom and India. "Motives. Research in the Universe, Own Theories and Field Torsion Program" conference. [60].

As an astronomer, I see the galaxies, the outer bodies, and the types of stars, and sometimes it is difficult for me to interpret what I see because of the distance from the outer body. But I realise that being a mathematician is very powerful and that only paper, pencil and great creativity and imagination can make a person travel through the universe, understanding and 'observing' all the phenomena that exist and happen.

3.3. A Walk in the Thoughts of Dr Francisco Bulnes

I am David Cote-Sánchez. I am an industrial engineer dedicated to teaching engineering and mathematics at the professional and degree level. I have had the opportunity to talk extensively to my friend, Dr Francisco Bulnes. I have been on one of his very interesting courses, *Guía de Prospectiva y Calidad de la Investigación Científica y Tecnológica Internacional*[5], where he proposes a research method to create a mathematical theory of nanotechnology based on the fundamental concept of quantum field intentionality. He described the finite actions of a field on matter and space and developed methods that define, determine and use as a basis the concept of the 'intentional action' of a field. This is so that its ordered effects realise the organised transformation of space and matter using integrals. As Dr Francisco Bulnes explains, human survival requires a technological future for the development of new technologies born from the designs that nature provides, and this is where it is necessary to understand and handle small, invisible objects, which requires a visionary and accurate method of science. This can only come from mathematics

We will begin this trip in the thoughts of Dr Bulnes to witness how mathematics, quantum mechanics, the curvature of the universe, electromagnetism, and levitation, predominate in his investigations.

In this narration, I have summarised the investigations on the previous topics. We will start our journey with his article entitled "Electromagnetic Gauges and Maxwell Lagrangians Applied to the Determination of Curvature in the Space-Time and their Applications" [25], in which he describes electromagnetic gauges and Maxwell's Lagrangian as being very important for the detection, metricity, sensing and creating of censorship conditions for detecting singularities such as supermassive objects or objects of immense gravity, such as black holes, white holes or other objects that can be undetectable without these fine gauges. This makes me think and imagine an environment like in the film *Star Wars*, where light

[5] "Prospective Guide and Quality of International Scientific and Technological Research".

rays, speed gauges for teleportation and direct interaction with the universe require devices based on these gauges, their sensors and transducers of energy: [25]

> finite actions of electromagnetic fields in Hamiltonian regime and use vector bundles of geodesic in movement of the charges with a shape operator (connection) that measures the curvature of a geometrical space on these geodesic (using the light caused from these points (charges) acting with the infinite null of gravitational field (background)) we can establish a model of the curvature through gauges inside the electromagnetic context (Bulnes 2012).

As we can see, this paragraph takes us into the quantised version of curvature and the effect of torsion in space-time. I hope the reader will be excited to imagine a space of electro-dynamic configuration to describe different interactions of the field in the universe related to gravity, with trajectories not seen before. Now, getting deeper, let us witness that intensity when moving into the world of mathematics of Dr Bulnes, in this same paper, when he suggests [25]:

> proposing and designing one detector of curvature using a cosmic censor of the space-time developed through distortional 3-dimensional sphere. Some technological applications of the used methods are exhibited.

Dr Bulnes explains how in continuous transmission media, there is a limitation that depends on the macroscopic characteristics of space-time, which causes electromagnetic fields to suffer refraction and reflection due to the substance of the media itself. His research proposes that to eliminate this direct action of the media in the transmission of signals, our concept of communication must not depend on the characteristics of the media.

In gravitation, there are field observables in an Einstein space (such as the curvature and torsion of space-time due to gravitational effects) that, in the electromagnetic case, the electromagnetic dispersion is presented as a distortion effect on space, where the waves act as distortion waves. In both cases, and here is one of the greatest contributions of this work, we can use the energy of these field observables to reverse this distortion effect, either by gravity or electromagnetism. Thus, Dr Bulnes invented a concept of energy curvature in the case of gravitation, but this extends to the energy of all field observables.

He begins his mathematical thinking journey by identifying those causal factors that he calls singularities in a complex Riemannian manifold[6], and that they are the dispersion in continuous media and interference due to the appearance of other waves, signals, contamination by free electrons, molecular absorption of the air, as well as defective routings. Secondary modes include optical fibre, gravity, radio sources, multi-radiative sources, and background. For this, he proposes that the transmission of the signal must have an invariance at the quantum level. My mind imagines waves moving in the universe without being deflected from their trajectories and that they reach their receiver completely free of any external action without undergoing any modification. It takes us into an invisible world. Let us continue with the paper "Electromagnetic Gauges and Maxwell Lagrangians Applied to the Determination of Curvature in Space-Time and their Applications", which has intrigued, or rather, fascinated me, which is why I return to it.

First of all, I documented what these 'gauges' are, and I found that in mechanics, they are called strain gauges, with the following characteristics:

- These strain gauges are sensors whose resistance varies with the applied force.
- These sensors convert force, pressure, tension, weight, etc., into electrical resistance, which can be measured through energy devices.
- These sensors are the most important elements in designing pressure transducers and load cells. However, these can be extended to other transduction processes. For example, in classic mechanics, the correct use of these gauges is in measuring forces and deformations as the most important tools in engineering or construction.

I must admit that as an engineer, I limit myself to my engineering vision of these gauges. However, I know that this has a larger connotation and scope. I observe how Dr Bulnes' thoughts are directed at analysing the current use of measuring forces in the universe through other forces (the gauges) invariant in the universe and how he wants to take advantage of this to perform experiments physically in a laboratory.

As I continued reading, I imagined his team concluding that these gauges are necessary because they monitor the deformations suffered by the mechanical components and that, in the most extreme case possible, ruptures in the said components can be detected in time, for example, in the

[6] The Kähler manifold models the universe. To involve microscopic aspects of the universe, it is necessary to introduce singularities in said variety.

forecast of ruptures and dislocation of materials and mechanical components. Although as I have managed to glimpse, I think this goes beyond simple mechanics. I will continue using the facsimile of classical mechanics because I consider that even at this level, the research on gauges is of great importance since it can be verified that the manufacture and miniaturisation of strain gauges have significantly advanced, impacting these transducers in different fields of mechanics, providing an improvement in the control of deformations. This research by Dr Bulnes has required an update on advances in engineering, and to my great surprise, I found research by Dr Francisco Bulnes, along with his Russian colleagues, on materials science and the determination of the exceptional qualities of metals through quasi-relaxation [61].

I stopped at this part since this topic is very exciting. Later I investigated the Lagrangian in modern physics, defining it as the most fundamental operator that describes a physical system relating movement with force and energy that is necessary to explore the mechanics in alternative physical systems to coordinate systems, such as polar, cylindrical, and spherical coordinates, or even the generalised coordinates for which Newtonian mechanics is not suitable.

CHAPTER 4

THE GOVERNMENT, SCIENCE AND EDUCATION: THE TRAGIC CASE OF MEXICO

4.1. Education as a currency of political interests

In a developing country like Mexico, education has always been under the shadow of the country's political and economic fluctuations. This should not be the case since education is supposed to lead to a greater consciousness based on knowledge for a balanced, structured and happy society, which is also physically, mentally and spiritually healthy. Without a good education, a society will never achieve this.

Dr Francisco Bulnes has made us see all the evils that deficiencies in education can bring, taking Mexico as an example, where he tells us that the Mexican educational system has been stagnant and subjugated for more than 60 years by the low-level policy of the Mexican state; that the Mexican state intends not to educate its people to maintain power. This intention has been enforced by persecution, murders, accusations, intimidation, and abuses in the educational sector, such as very low salaries and precarious working conditions. How can a politician unrelated to education (without academic preparation or a career related to teaching) manage and govern a university, technical college, educational council, or government secretariat? Dr Bulnes told us that this happens and is very common in Mexico and that this and other vices keep the quality of education far below standard. He even made us see that something very paradoxical was happening in Mexico. On the one hand, Mexico has a very low basic level of education (in elementary and upper secondary education, even in some cases, at the professional level). On the other hand, at the scientific level, Mexico has great scientists and academics with postgraduate degrees. What is happening here?

For the bad governments, populist or non-populist, that Mexico has had for more than 70 years, including the last governments, education and science are not their priorities.

Curiously, Dr Bulnes told us that it often depended on the perception and concept that a politician (as a singular individual) had about education. He remembered some directors of a technology company who were aware of the importance of science and education, putting aside their political pejoratives and political ideology. These are individual cases where the person acts according to their perception of the importance of education and science and their principles and educational level.

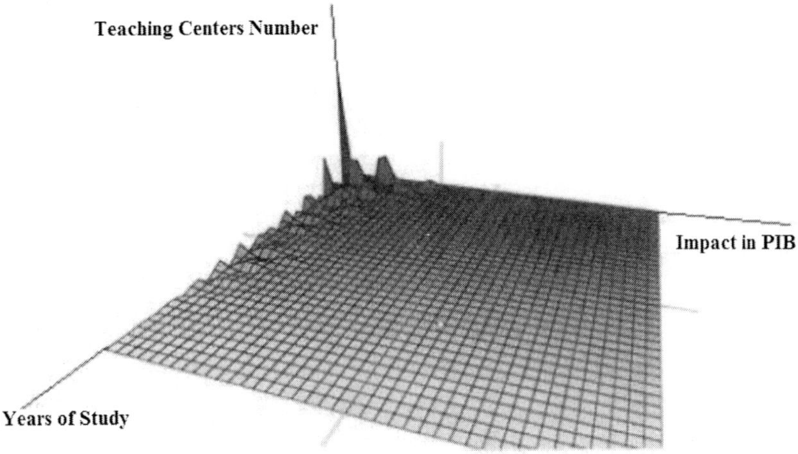

Figure 31. Impact of the PIB on education in Mexico [44].

In basic education, Mexican students fail tests. In scientific research, where only a few students bring the skills from home required to reach an advanced level of science, there are notable scientists and brilliant academics comparable to those in developed countries or with a great scientific tradition. Many of the students will not be scientists, but they will be professionals, technicians and managers, in which case the comprehensive development of a country will depend on their training (see Figure 31).

However, apart from that, the basic and upper secondary education level is below international standards. Why? What is this phenomenon due to? Why, specifically in Mexico? The last question has to do with the fact that countries with an almost non-existent GDP, such as Cuba, Haiti and Bolivia, have better basic education, and not only that, better future projections of GDP (see Figure 32) due to the implementation of improvements for their development [62] (for example, in the cases of Cuba and Bolivia) (see Figure 32).

Curiously, the Mexican educational system is one of the largest in the world; however, its efficacy and efficiency are very low. Also, the Mexican educational system has failed to engage the indigenous and minority populations comprehensively.

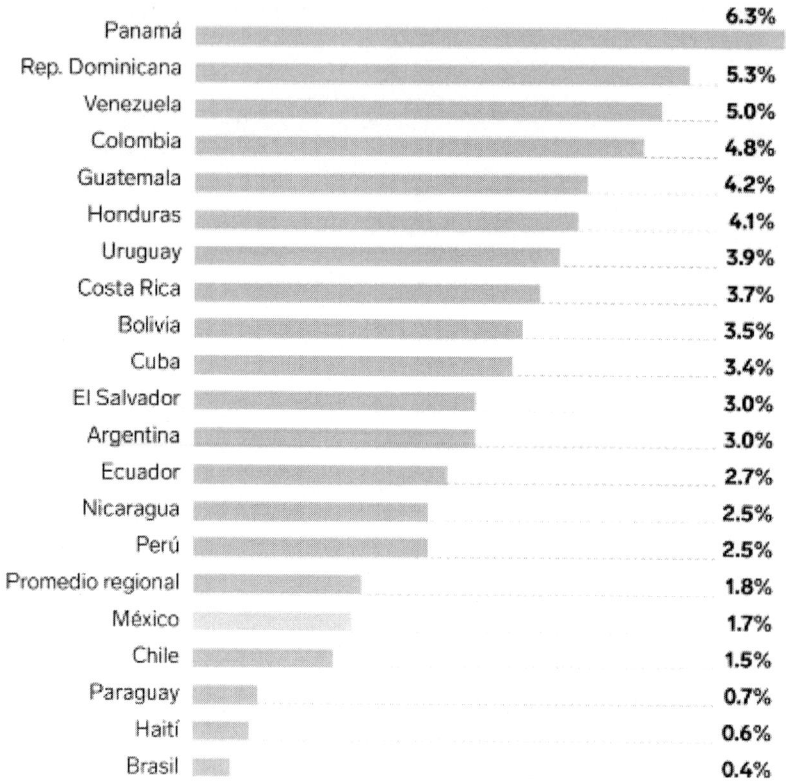

Figure 32. GDP growth forecast (%) according to CEPAL [62, 63][1].

Institutes and universities specialising in education, for example, the UPN and the Superior Normal School, etc., have not been able to raise the academic quality at the basic and upper-middle levels. In some cases, the professional level is affected due to the lack of educational specialisation, with no appropriate selection process. There is no comprehensive teacher training programme.

[1] CEPAL (Comisión Económica para América Latina y el Caribe).

What explains all this is the politics that prevails and permeates educational matters in Mexico. Dr Bulnes has mentioned that many technology universities are in charge of technical education through engineering programmes and technical careers, directly impacting the industrial sector and, therefore, internal and external economic growth. However, the universities are seen as booty for the politicians and political parties who plunder their resources, facilities and infrastructure for their political lobbying and popular programs far removed from education. Many Latin American countries possibly have very similar problems. However, this is more pronounced in the Mexican case due to the government's power structure and the organisations that succeed it, in addition to so many sub-secretaries and departments of the corresponding Ministry of Mexican Education (in this case, the SEP). In many cases, these do not act independently and objectively, and most are 'dead' departments, which increases the bureaucracy for what is required in the different education programmes.

However, Dr Francisco Bulnes tells us that the causes can even come from abroad, where organisations such as the OECD, IMF and the World Bank establish productivity strategies for a country in order to create productive individuals, not necessarily in the intellectual sense, but in the economic sense, giving priority to the economy more than academia. This, in principle, could not be so bad since it seeks to increase the country's GDP and improve living conditions for its inhabitants. However, the real intentions of the World Bank and the International Monetary Fund are to establish aggressive policies for the countries in order to lend them money with high interest rates (at the end of the day, it is a business) and each country, in order to comply with these expenses, raises the internal interest rates and the education sector must be productive, disdaining the pure sciences and art in all its manifestations. In principle, the pure sciences are the basis of knowledge, and education is the only way to have a sufficient background of essential knowledge.

An academic's social and economic accommodation is arbitrary and unfair in Mexico. Many academics, even with PhD degrees, barely have enough to pay the mortgage on their home and afford necessities. Meanwhile, ignorant, corrupt and poorly educated politicians keep all the money and make fortunes jumping from one party to another, having no ideology or commitment to the people.

Another identifiable vice is the facilitation without any intellectual effort of some pseudo-academicians who form brotherhoods with elements of the government, education leaders, and education unions to obtain academic degrees, stipends and support for projects that have not been fully evaluated. The latter happens very often with pseudo-academics who want work positions, unearned research status, and even ask for office space that has not been earned since their work lacks the minimum quality, capacity and inventiveness required.

Mexico, in recent years and with the degradation of the educational system and corruption in the corresponding government institutions responsible for promoting scientific research and quality academic work, has an exponential growth of these vices, endangering the country's development. As previously mentioned, the development of a country is based on education and, by extension, scientific and technological research.

4.2. External Interests in Mexican Education

As mentioned, education in the Mexican case supports various group interests, parties, the government, and even foreign policies implemented by the World Bank and the International Monetary Fund.

However, we can also add the interests of developing countries that want to make the country a captive market dependent on technology and educational models, which is very far from the Mexican reality, which they do through powerful television advertising. We are sold the 'American dream,' full of well-being, abundant food, and great comfort from 'conveniences' and a standard of living defined as a great way of life. In principle, what is "quality of life?" Dr Bulnes states, "Quality of life is the state of life that every human being can achieve through inner development to achieve full happiness and well-being based on mental, emotional and physical health." This definition does not explicitly mention money as a satisfier.

However, it is implicit since the person's inner development will make them obtain the necessary resources to satisfy all their needs, although they will always do it with measure and equity. To obtain this inner state that leads us to a good quality of life, an individual's education is required to achieve this higher consciousness to access and create wealth.

Consider how Thomas Jefferson, John Adams, Benjamin Franklin, and George Washington designed the first Declaration of Independence. We observe that they consider happiness a universal right where people, upon reaching this state, may have the ability to transfer that well-being to others, creating a healthy social fabric, that is, a healthy society. This document established no interventionism, commercial or intellectual invitation from the US. However, in practice and due to the economic interests and politicians, many commercial influences and education is governed by these interests from Washington politics, where the United Nations (UN) must support the rich countries' politics and policies. Then the educational goals are also influenced at a primordial level, decreasing the chance of education being independent.

However, some towns have home education, and this can help lay the foundations for an honest, true person interested in knowledge, with principles about honour, the common good, and peace as a basis for all the values of democracy, constructive dialogue based on reason, etc., forged at home. These values are independent of the country or region; they are immutable since they should never change and are invariant of the people and their customs since they do not depend on the particular worldview of a country, culture or society. It is the only thing that can save a people from being annihilated, stolen or conquered. The principles do not depend on money, although money is a means to have physical and material satisfaction (food, shelter, clothes) – it will never be food for the spirit. The food of the spirit is truth, love, knowledge and the desire to be better as a human.

4.3. His trips to Cuba

Francisco Bulnes was invited to Cuba many times. The first time was for a post-doctoral stay during 2006–2007, covering various academic activities and his research. His post-doctoral research consisted of applications of infinite Lie theory in information and systems theory [64] (see Figure 33). Dr Bulnes taught as a visiting professor at the University of Informatics Sciences, Havana, Cuba. For the course on algorithm theory for the master's degree in applied informatics, I carried out some collaboration agreements between the University of Informatics Sciences and the Institute of Mathematics of UNAM. He met various sectors of the University of Information Sciences and gave some presentations on research carried out with applications to systems.

During the post-doctorate year in Cuba, Francisco Bulnes was approached by the José Antonio Echeverría Higher Polytechnic Institute in Havana, Cuba. There he began working on his FET and disclosed, through a plenary conference, the principles of this theory on engineering development [66].

A very flattering comment about his theory was received from another Mexican scientist, Dr Merchan, who had been accompanying his son Dr Emmanuel Alejandro Merchán Cruz, to an international event, the 14th Scientific Convention on Engineering and Architecture (CCIA 14) in Havana, Cuba, where the most prominent Latin American scientists met and presented their research and results at various congress forums [66, 67].

Verbatim, he said that the theory proposed by Dr Bulnes could not perhaps be understood with all its value due to the abstractness and rigour with which this theory was presented, but that future generations, possibly twenty or fifty years from now, would value it as a fundamental theory about engineering and its epistemology.

Figure 33. The university published the course on algorithm theory taught by Dr Francisco Bulnes, and some results obtained in his research during his stay in Havana are included. A book on the postgraduate course taught by Dr Francisco Bulnes during 2006–2007 was published as part of the activities to be carried out during his post-doctoral stay. Dr Carlos Novo and Ms C Iván Mallea (postgraduate director at UCI) were in charge of the book's editing and publishing [64, 65].

For Francisco Bulnes, Cuba was also very intriguing for the social struggles it represents and the international struggle against the power of money. He has always admired CdTe Ernesto 'Che' Guevara and Fidel Castro Ruz, whom he always considered exemplary men, full of courage and determination. So when Dr Francisco Bulnes went to Cuba for the first time, it was a great event for him (also an important event for Cubans), full of feelings of all kinds, since Dr Bulnes had also been a great admirer of the Cuban revolution with its victory in 1959, the story of José Martí, and his fight against the international blockade which by then had lasted more than 40 years.

He saw Cuban cinema, read several modern literary authors, and generally liked to read about the anecdotes of Che and Fidel Castro when they were in the guerrilla war trying to lead the armed revolution of Cuba towards victory.

Figure 34. Daisy Granados, the face of Cuba for Dr Francisco Bulnes.

He tells us that he had the face of Daisy Granados in his mind when on his way to Cuba (see Figure 34), and for him, Cuba was the face of Daisy Granados and vice versa because he had seen her in Cuban films [68].

During his first stay in Cuba, he made great friends with a highly respected researcher in Cuba, Dr Carlos Eulalio Novo Soto[2], who immediately assisted Dr Francisco Bulnes during his academic stay as visiting professor at the University of Information Sciences from Havana, Cuba (see Figure 35).

---

[2] A Doctor of Science in Mechanical Engineering and top-level researcher at the José Antonio Echeverría Higher Polytechnic Institute, at CUJAE (Engineering Research Unit). A researcher at the University of Informatics Sciences, considered one of the great doctors of science trainers in mechanical engineering in Cuba.

Figure 35. Here we see Dr Francisco Bulnes at a dinner inside the rectory of the UCI, which was offered to welcome him as a great academic who would share his knowledge. Beatríz Aragón, director of international relations, appears in the foreground. Behind her, the postgraduate director of the UCI, MsC. Iván Pérez Mallea. Behind him, Dr Carlos Novo Soto, flagship researcher of the UCI. On the other side, the Vice-Rector Rosa María Vázquez Cedeño, the Rector of the UCI, and finally, Lic. Rosa Roqueful-Helster, sub-director of the Logistics department of UCI. In the middle and at the head of the table is Dr Francisco Bulnes, honoured with dinner [69].

Figure 36. Dr Francisco Bulnes, flanked by MsC. Iván Pérez Mallea (another great friend) and Dr Carlos Novo in the graduate section of the University [69].

Figure 37. Dr Carlos Novo, a great friend of Dr Francisco Bulnes. Here he is receiving a doctorate honoris causa from ISPJAE.

Dr Novo frequently communicated with Dr Bulnes through email and regular mail. Dr Francisco Bulnes, as the main chairman of an international event called Appliedmath, invited Dr Novo to participate, along with other scientists from other countries (see Figures 36 and 37). Dr Francisco Bulnes undertook some projects with Dr Carlos Novo, and their friendship was important in realising agreements between Cuban universities and the Mathematics Institute of UNAM, where Dr Bulnes worked in those years.

Figure 38. An advertisement for the 51st anniversary of the Cuban revolution, seen during a walk through Havana showing Fidel Castro Ruz and Camilo Cienfuegos Gorriarán, heroes of the revolution.

When Dr Bulnes went to Cuba, Dr Novo received him with great appreciation, and they drank some mojitos in the centre of Havana (the old part of the city) (Figure 39). Dr Novo learned of Dr Bulnes's admiration for Che, and Dr Novo gave him a book written by Che himself about his last campaign in Bolivia [70].

Figure 39. With Yudeisy Pérez and Dr Novo drinking mojitos in Havana, Cuba. Appearing from left to right Dr Novo, Yudeisy and Dr Bulnes [69].

Figure 40. Dr Bulnes in the classroom, teaching the postgraduate course, "The Theory of Algorithms" [69].

Figure 41. On another trip to Cuba in 2008, Dr Bulnes, an admirer of the Cuban revolution and Commander Ernesto 'Che' Guevara asked to be photographed in Revolution Square, in front of the building with the effigy of Che on it [69].

The trips to Cuba and his approach to the Ministry of Higher Education showed Francisco Bulnes how admirable Cuba's struggle was for wanting to be a sovereign country and that although besieged by the blockade imposed by the US and its allies, Cuba had all the infrastructure for education, and was comparable to European countries (Figures 35, 36 and 40).

He was invited as an important delegate from Mexico to verify the facilities of the University of Informatics Sciences of Havana, Cuba, one of the many universities in Cuba specialising in computer sciences, systems and informatics (Figure 41 and Figure 42).

It was unthinkable that such a small island, barely the size of the state of Veracruz in Mexico, would have specialised universities with modern facilities and state-of-the-art laboratories. It must be said that much of this infrastructure belonged to some European countries, such as Spain, which had many investments in Cuba and Russia (at the time of the USSR[3]). Further, Dr Bulnes found that Cuba invested almost everything it earned in education.

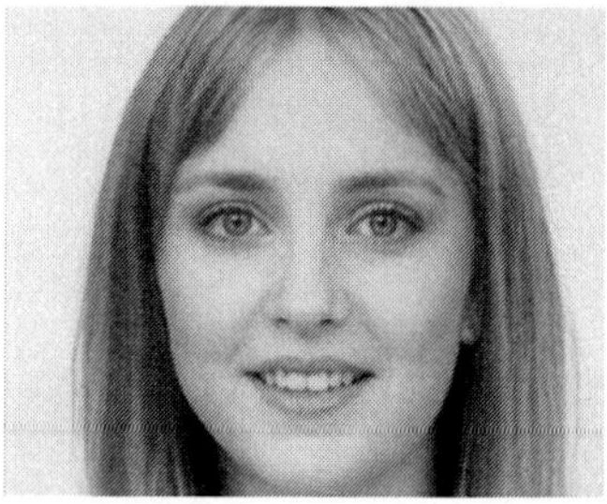

Figure 42. Lic. Rosa Roqueful-Helster, sub-director of the Logistics Department of the UCI, invited the academicians. She accompanied Ms C. Ivan Pérez Mallea (postgraduate degree sub-director at the UCI) and Dr Francisco Bulnes on the trip that the UCI offered him to Varadero, Cuba. She brought her son Yudeisy [69].

His experiences with the people of Cuba were very pleasant. He visited almost the entire island from Havana to Santiago de Cuba, Varadero to Viñales de Pinar del Rio. Cuba has a diversity of natural scenarios, ranging from the mountainous part of Pinar del Rio to the beaches of Varadero (see Figure 43), where the sand is white. The sea is light blue, changing to shades of turquoise.

[3] USSR—the Union of Soviet Socialist Republics, 1922–1991.

Figure 43. Varadero, Cuba [69].

Figure 44. This was part of the agreement signed and published in the UCI newspaper. It includes joint research programmes and areas with the UNAM Institute of Mathematics, which Dr Francisco Bulnes represented. The president was Dr Bulnes. This programme was interrupted by the six-year term change in Mexico. However, in Cuba, some of these programme plans were followed [71].

Finally, it has been said that Castro's government was dictatorial and removed many freedoms. However, I have to say that many times after the 59th anniversary of the revolution, Cuba was in danger of losing its autonomy as a country. The political, military and economic siege it suffered for more than 50 years was a benchmark that forced the Cuban government to close ranks and ensure that the values of the revolution for which they fought and conquered were not lost. I believe that Cuba must continue to care for these values, add new ones, open up to the market and take steps towards democracy.

Figure 45. Notice on the intranet and newspaper of the university in 2007 [71].

Fidel Castro's government was heroic, and it was what Cuba needed at that time to be a sovereign country; Cuba's resources belonged to the Cubans themselves, and the country's wealth was returned to its inhabitants, with education as the primary objective of its government. Now the time has come to take the next step, strengthening its economy with a free market, open socialism, and economic opportunities in the country's interior.

The academic activities carried out by Dr Bulnes were transcendent and changed some aspects of the UCI postgraduate course (see Figure 44 and Figure 45).

4.4. The Proposals for Improvements in Education in Mexico

Dr Bulnes' trips to Cuba and his talks with its Ministry of Higher Education led him to identify the root of the problem in education in Mexico. This lies in the lack of a common objective of superior growth based on the knowledge and development of science. Dr Bulnes commented to his Cuban colleagues, "You do not have the problem of many bosses, that is, a bureaucratic apparatus that delays and destroys the initiative... you only have one direction... because you only obey one boss and initiatives in education are taken quickly."

In this bureaucracy, politics, illegitimate economic interests for education, freeloaders, syndicalism, and other problems derived from diverse interests corrupt and divert the purposes of education (see Figure 46). One of the most identifiable problems in Mexico is corruption; as was mentioned before, basic positions for teachers are sold, and the resources destined for education are diverted to populist programmes aimed at collecting votes. But the most serious problem is that the Mexican people lack historical memory; the family does not promote the fundamental principles of life. Cognitive skills are not developed by the family or shared between parents and children, and above all, the majority of the Mexican people live with economic shortages and underemployment, which make them more concerned with dealing with basic survival and the long working hours, so they do not have the time to develop an interest in knowledge and education.

In Mexico, we do not promote science, art and music, so achieving scientific and technological developments in primary or high schools or universities becomes impossible. [43, 44, 73].

Figure 46. Dr Bulnes receive recognition from the Iberoamerican Doctoral Cloister of Mexico for his educational leadership. He is accompanied in receiving this award by his friends and colleagues from the National Polytechnic Institute (IPN), Dr Emmanuel Merchán Cruz and Dr David Fragoso. He was recognised for his contributions to new concepts in education [72].

4.5. His Meeting with the US: an Important Academic Passport

Dr Bulnes travelled to the US after receiving invitations to participate in several symposia, international conferences, and meetings. As a member of ASME (the American Society of Mechanical Engineers) for more than 11 years, he received many invitations to participate as a lecturer, speaker, and organising committee member in the different sections of international conferences on very specific themes. Dr Francisco Bulnes frequently participated in the section on emerging technologies [74, 75], where at various times, he gave expositions on photonics, advanced electronics, advanced mechanics and advanced dynamics through his synergic integral operators (see Figure 47) and later, he lectured on the fundamental mathematical principles/theory of nanotechnology. He visited all 50 states in America and gave some interesting talks on nanotechnology-related themes, such as nanomedicine and his research in nanomaterials. Dr Bulnes also visited Vancouver, British Columbia, in Canada in 2010 [76].

Figure 47. Dr Bulnes gave his talks on nanomedicine using his synergic integral operator theory to explain methods of cure at the atomic level using special bosons. This talk was given in Denver, Colorado, US, in 2011 [77].

All these trips and presentations in the US earned him the fame of a great nanotechnologist and scientist for the foundation of possible future technologies, making aspects of quantum electronics, photonics and spintronics possible.

On one of those trips, he received an invitation from a publishing house in the US to be editor-in-chief and the founder of the *Journal of Photonics and Spintronics*. Dr Bulnes accepted and acquired the ISSN and eISSN for the journal and achieved its indexing in the four most popular scientific databases on academic works. Dr Bulnes published more than 24 issues in six years. In the compilations made in the six volumes [80], many works have been published that could provide the bases and principles of future technology governed by particles, fields and waves. Dr Bulnes organised a small event at TESCHA[4] with top-level researchers in areas related to photonics and spintronics (see Figure 48).

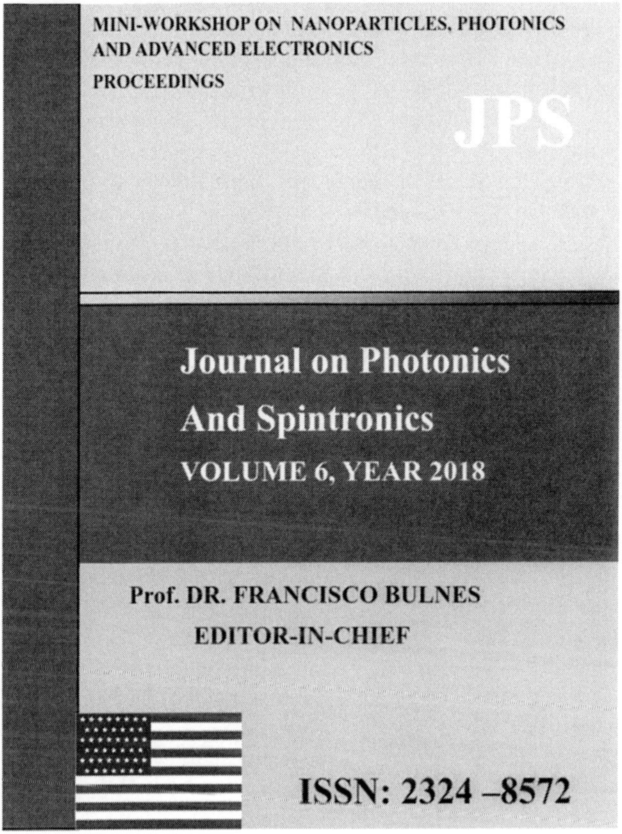

Figure 48. Proceedings of the mini-workshop on nanoparticles, photonics and advanced electronics. This was the special issue of the *Journal on Photonics and Spintronics.* Dr Francisco Bulnes has been editor-in-chief of the US edition for eight years [78]. The journal contests several advanced technological secrets [79]. These secrets coded through field theories will be revealed in the next few years.

The quantum mechanisms of fermions, a fundamental piece for field interactions, insulators as windows of the universe, caloritronics for the use of temperature changes to generate energy, the principles of quantum

---

[4] Tecnológico de Estudios Superiores Chalco.

communication, the core of a superconductivity reactor, particle and anti-particle interactions, Majorana fermions, spinor technologies, hermeneutics, possible devices in nanomedicine, and the microscopic aspects of the universe are all explained. These volumes[5] were produced directly by the government of the US because they suggest interesting future technologies in various ambits and items of technological interest for the survival of humanity, and the US saw great potential in this research and the content of these volumes [80].

After a few years, Dr Bulnes was commissioned to undertake several research tasks in the US and was invited to join several international committees of science and scientific ethics. He was also invited to join the American Nano Society as a full member[6] and to write articles in several scientific journals. Dr Bulnes established contact with some American academicians and scientists in the nanotechnology and mechanical engineering branches.

His research on electromagnetic propulsion brought him an invitation from NASA and other organisations dedicated to the research in astronautics and aerodynamics engineering to give a talk as a speaker on the advances of astronautics and flight vehicle research (Figure 49).

Figure 49. Promotional presentations given by experts in astronautics and aeronautical engineering. Dr Bulnes appears among the panellists of invited speakers. This event took place in Orlando, Fl, USA, in 2016. That year they had the most expertise in the world on these themes [81].

---

[5] Six volumes in eight years were produced by the US.

[6] Status that is only granted for merits in the field of nanotechnology.

4.6. His trips to Central Europe: Germany, Poland, Slovakia, Serbia, Czech Republic, Croatia, Monte Negro, Romania, and Lithuania.

Dr Bulnes was invited to give different talks and lectures on the pure mathematical aspects of his research. He travelled to Germany frequently and into Central Europe by train. He visited the Czech Republic, Croatia, Monte Negro, and Romania and participated in conferences and international meetings in mathematics, geometry, topology and global analysis.

He participated in FSDONA 2011, organised by Dr Dorothy Haroske, a reputed mathematician from the Friedrich-Schiller-Universität, Jena.

He explained and disclosed his work *Cohomology of Moduli Spaces in Differential Operators Classification to the Field Theory II* [82], which is considered high level. See (Figures 50 & 52).

On another trip to Germany, Dr Bulnes gave talks, invited several European mathematicians and academicians to his seminars, and travelled around Central Europe (see Figures 51, 52 A and B).

Figure 50. Dr Bulnes is pictured with Dr Dorothy Haroske, a distinguished German mathematician who researches function theory on infinite dimensional spaces, differential operators and non-linear analysis, and she directs the research group at the Friedrich-Schiller-Universität, Jena. For many years she has directed the international event FSDONA[7]. Dr Bulnes participated in the lecture "*Cohomology of Moduli Spaces in Differential Operators Classification to the Field Theory II*". Here we can see two world-renowned mathematical leaders.

[7] FSDONA (Function Spaces, Differential Operators and Non-Linear Analysis).

Figure 51. A little town near Szczecin in Poland. Dr Bulnes found mathematical colleagues here to discuss some ideas about global analysis research [83].

A B

Picture 52. A). On another trip to Germany, after his talk on his research, Dr Bulnes walked into Jena with his German colleague Dr Leopold, Hans-Gerd. B). The logo of the University of Jena [84].

In Slovakia, Dr Bulnes befriended the most renowned Czech mathematician, Dr Demeter Krupka, who lent him a mathematics monograph of his own, his *Introduction to Global Variational Geometry*, something unusual from one mathematician to another since mathematicians do not usually lend their books out, considering them the most valuable things in their possession (see Figures 53 and 54).

Figure 53. Dr Demeter Krupka and Dr Francisco Bulnes, during the sessions of the 20th International Summer School on Global Analysis and its Applications: General Relativity 100 Years after Hilbert [85, 86].

Figure 54. In Slovakia, during a mathematics meeting in 2016. This meeting was dedicated to Einstein's theories, symplectic geometry and complex Riemannian manifold topology of low and high dimensions [86].

Dr Bulnes was awarded a doctorate honoris causa in Frankfurt on another trip to Germany in 2018. The decoration and appointment of the doctorate honoris causa in the philosophy of education took place at the castle of the Empress Victoria, originally called the Schloss Friedrichshof, in honour of her late husband, Emperor Frederick III (Friedrich III) (see Figure 55).

Picture 55. Dr Francisco Bulnes received the doctorate honoris causa in the philosophy of education and the Ambassador of Peace Award from the International Red Cross (IRC) in Frankfurt in 2018, as seen by the medals. The award gowns were made of fine silk and gold. He is shown here in one of the rooms of Kronberg Castle [87, 88].

Figure 56. The exterior of Kronberg Castle and a hallway.

Later he travelled to Serbia, Croatia and Montenegro, passing through Kosovo, where he gave a series of conferences on his latest results on derived categories and geometric analysis methods for moduli spaces (see Figure 57 and Figure 58).

He had a special reception in Belgrade, Serbia, which he returned to three more times. Dr Bulnes made friends with Dr Branko Dragovich, Dr Marko Vojinović and Dr Igor Salom.

Figure 57. In Serbia, during the 9th Mathematical Physics Meeting: School and Conference on Modern Mathematical Physics.

A year later, he would participate at the Institute of Physics Belgrade with the talk "Baryongenesis until dark matter: H-particles proliferation." Two talks were published in two different issues and volumes of the official proceedings of the Institute of Physics Belgrade.

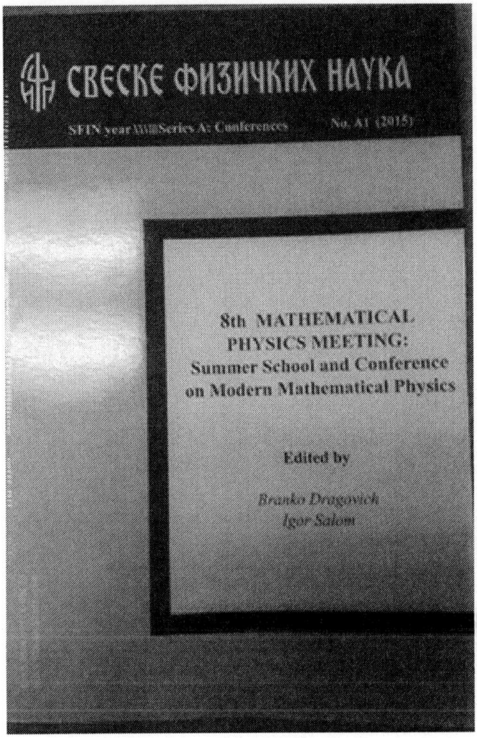

Figure 58. Official proceedings of the School and Conference on Modern Mathematical Physics. Edited, printed and published by the Institute of Physics Belgrade [84].

Dr Bulnes affirmed that he liked travelling to Eastern Europe; there is still an old flavour typical of traditional Europe, full of folklore and reminiscent of yesteryear (see Figure 59). As a respected and famous academic, several dances were performed for him at a reception.

Figure 59. Traditional dances as seen on the travels of Dr Bulnes through Eastern Europe [84].

One of the mathematical approaches that has fascinated Dr Francisco Bulnes for the study of geometry, dynamical systems, and the nature of microscopic phenomena for years is that of integral transforms: their evaluations and the invariants and methods that can be developed when you have the so-called singular integrals within equations, along with the solution for non-linear integral equations. In this respect, Dr Bulnes compiled integral equations with their problems and the development of numerical methods to solve them.

Dr Francisco Bulnes has a great friendship based on shared educational values with the CEO, Valeria Dos Santos-Reiser, from ODAEE (see Figures 60 A) and B), 61). Dr Santos-Reiser undertook projects to increase the quality of education in all universities in Latin America, Africa and poorer countries in Asia. As a German-Brazilian, she considers mathematics a perfect science for developing more cognitive abilities.

A)

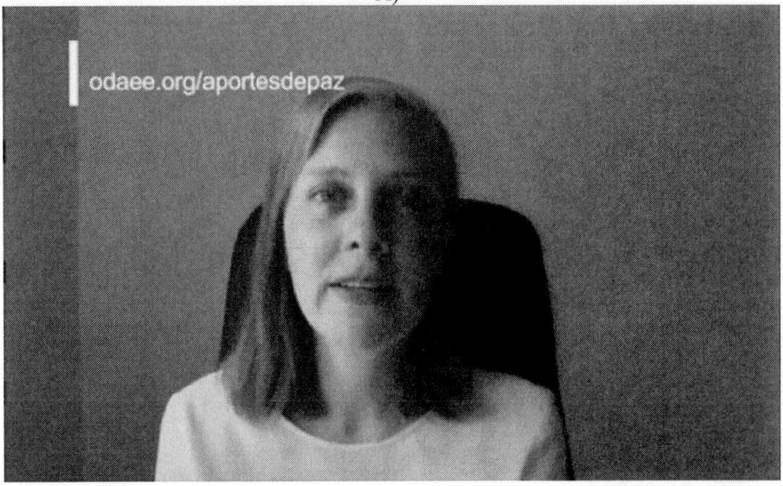

B)

Figure 60. A). Dr Bulnes is interviewed by Valeria dos Santos-Reiser, the CEO of ODAEE, with the lecture "La Verdadera Formación de la Educación: Personas hacia el Pensamiento Crítico" at three different events in three different countries. [89]. B) The CEO of ODAEE announcing the investiture of his doctorate honoris causa in Milán.

Figure 61. Dr Bulnes with Valeria dos Santos-Reiser, CEO of ODAEE, during an educational seminar.

One of the many concerns that Dr Bulnes has had is the development of functional analysis through all his lines of research that he has been developing with several of his Russian friends and colleagues: Dr Fominko from Ukraine, Dr Ivan Verkelov, Dr Rabinovich and Vladislav Kravchenko from Russia, (now living in Queretaro, Mexico and who works at CINVESTAV— The Center for Research and Advanced Studies of the National Polytechnic Institute).

Dr Bulnes, with Dr Verkelov and Dr Stropovsvky, has worked on these functional analysis lines: the representation theorems of functionals and operators, the Hahn-Banach theorem, integral equations, integration in infinite dimensional spaces, bounded and unbounded operator theories, Banach algebras, Sovolevand Orlicz[81]spaces, some generalisations in harmonics analysis, and symbol theory, etc. [27, 90, 91, 92].

[8] In mathematical analysis, and especially in real, harmonic analysis and functional analysis, an Orlicz space is a type of function space which generalizes the L^p, spaces. Like the L^p, spaces, they are Banach spaces.

Figure 62. Dr Bulnes was the academic book editor celebrating Hilbert, working on a book with special advanced topics on integral equations. In the first pages Dr Bulnes includes a phrase of Hilbert addressing the community of mathematicians in the world [93].

The themes of recent research are non-linear integral equations, the new numerical and adaptive methods of resolution of integral equations, the generalisation of Fredholm integral equations of the second kind, integral equations in time scales and spectral densities, operator theories to non-symmetric and symmetric kernels, extension problems to Banach algebras to kernels of integral equations, singular integral equations [94], special treatments to solve Fredholm integral equations of the first and second kind, non-degenerated kernels and symbols of integral equations, topological methods to resolve integral equations and representation problems of operators of integral equations (see Figure 62).

One of the new developments in non-linear integral equations is followed by the Hammerstein integral equations [95], which are written as:

$$\omega(t) + \int_a^b K(t,s) f\big(s, \omega(s)\big) ds = 0, \quad a \leq t \leq b, \qquad (7)$$

where $K(s,t)$, and $f(s,t)$, are given functions, while $\omega(t)$, is the unknown function. Hammerstein considers $K(s,t)$ the symmetric and positive Fredholm kernel. This last condition establishes that all their eigenvalues are positive. Thus, the function $f(s,t)$, is continuous and satisfies

$$|f(s,t)| \leq C_1|s| + C_2, \qquad (8)$$

where C_1, and C_2, are positive constants and C_1, is smaller than the first eigenvalue of the kernel $K(s,t)$, then the Hammerstein integral equation at least has one continuous solution. Consider certain observations on the non-decreasing function $f(s,t)$, on s, considering t, from the interval (a,b). Hammerstein's equation cannot have more than one solution. This property also holds if $f(s,t)$, satisfies the condition

$$|f(t, s_1) - f(t, s_2)| \leq C|s_1 - s_2|, \qquad (9)$$

where the positive constant C, is smaller than the first eigenvalue of the kernel $K(t,s)$. A solution to the Hammerstein equation may be constructed by successive approximation. On this point, many approximation methods are designed to solve these integral and other non-linear integral equations. Also interesting are the recent developments in Hammerstein-Volterra integral equations:

$$f(t) = \omega(t) + \int_a^\tau K(s,t) f\big(s, \omega(s)\big) ds, \quad 0 \leq t \leq 1, \qquad (10)$$

The aspect of the linear integral equation has been important for studying Volterra integral equations on a time scale, where the initial value problems with unbounded domains have more importance. Likewise, the development of the alternative form of a linear integral equation is given as follows:

$$f(t) = \omega(t) + \int_a^t B(t,s)\, \omega^\sigma(s) ds, \ \forall t \in I_T, \qquad (11)$$

where $B(t,s)$, is a kernel that comes from Banach algebra, and ω^σ arising naturally from changing dynamics problems, for example, economic dynamics. Some aspects of their perspective can be extended to the non-linear case.

Dr Bulnes used a Volterra-Hammerstein integral equation to determine and express the dynamic analysis of a mean curvature energy illumination sensor (see Figure 63). Here the sensor must start the response signal for the luminous efficiency described for the mean foreseen behaviour given for the integral equation:

$$\eta(t) = h(0)\lambda(t) + \int_{0}^{\infty} h(t-\tau)\lambda(t)dt, \qquad (12)$$

being $\lambda(t)$, the luminous efficiency and $h(t) = V_{in} \exp\left\{-\left(\frac{\lambda}{R(\phi)}\right)\right\}$, the response signal of the sensor. The integral equation represents the functioning of the sensor submitted to luminous efficiency $\lambda(t)$, which will be relevant for detecting and measuring the illuminating curvature energy. Dr Bulnes obtained images of the spectra bandwidth of the mean curvature spectra and the dimensionless value λ [96] (see Figure 63 B). Considering certain physical facts, the Volterra-Hammerstein integral equation is reduced to a simple Volterra integral equation.

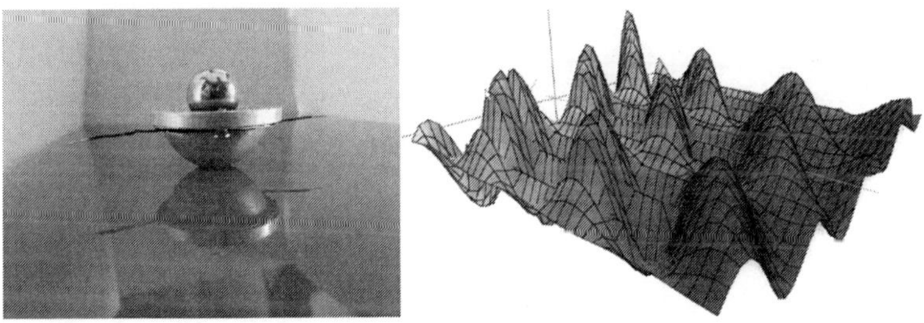

Figure 63. A). A captor device of average light in a three-dimensional space. B). Mean curvature spectra. The range $\kappa_2 \geq H(\omega_1, \omega_2) \geq \kappa_1$, is taken on the Y-axis [96].

4.7. More Recognition for His Impressive Academic Trajectory

At the end of 2017, Dr Bulnes announced his election and accreditation to receive the doctorate honoris causa from Selinus University (Figure 64), headquartered in Italy and other European countries.

A)

B)

Figure 64. A). District of Ragusa, Sicilia, where Dr Bulnes received the award. B). The doctorate honoris causa announcement inside Selinus University in Ragusa.

Interviewed by *AMAQUEME*, a local newspaper of great importance in the eastern zone of CDMX and the State of Mexico (the central zone of Mexico), he announced that the honorary doctorate that had been given to him on the 17th of February 2018 [97], and was for his 25 years long academic and scientific career in the field of mathematics, which was recognised worldwide (Figure 65).

Figure 65. Dr Francisco Bulnes, in an official photo for the university where he works as a researcher [97, 98], shows the doctorate honoris causa certificate.

In Italy, the award ceremony was attended by the education minister of the Italian government in the district of Ragusa, Sicilia and representatives of the Commonwealth of Dominica (Figure 64 and Figure 65).

The public notary licensed in Italy, Dr María Schembari, granted the award certification from the Ragusa and Modica College of Notaries to authenticate signatures in said award [97, 98].
Dr Bulnes commented in the media.

> Several professors from that country have seen my publications on the internet or by other means and have invited me to collaborate at Selinus University; by giving me that doctorate they made me an associate professor at that university in the area of engineering [98].

Other honorary doctorates were awarded to Dr Bulnes by other universities, NGOs and GOs.

CHAPTER 5

EXPERTISE AND AUTHORITY IN SEVERAL MATHEMATICS AREAS, THEORETICAL PHYSICS AND NANOTECHNOLOGY: A UNIVERSAL MATHEMATICIAN

5.1. Authority and Leadership

I am Dr Helen Thrift, and I will elaborate on some aspects of the academic contributions and results in mathematics and other sciences that Dr Francisco Bulnes has obtained and carried out throughout his academic career so far. Of the many things that Dr Bulnes learned from his mentor, guide, friend and doctoral thesis director was studying modern mathematical topics, theories and multi-disciplinary mathematical research on difficult topics. He recalls his great teacher Dr Recillas saying: "You must study the modern themes and theories of mathematics that are studied in the great mathematical research centres and universities of the world". For this reason, Dr Bulnes states that Dr Félix Recillas-Juárez was a universal mathematician of international stature whose work and research were consistent with the research problems of the world's great mathematics institutes. As mentioned, Dr Recillas was recognised by the best mathematicians of his generation, and even on his 80th birthday, many of these great mathematicians came from Europe, Asia and North America to pay tribute to him. This encouraged his students to keep up to date with modern and highly advanced theories that were developing alongside training with Dr Recillas.

The seminar, the *Theory of Representations of Real Reductive Lie groups* celebrated since the end of the 90s and the first decade of 2000, was a cornerstone in the formation of world issues such as motive cohomology, the infinite Lie theory, inverse cohomology, tensor derivative categories, complex varieties, the geometric problem of Langlands, complex singularities, Morse theory, global analysis, and the cohomological theory of Vogan-Zuckerman, among other topics. In all these subjects, Francisco Bulnes was trained and developed his own results, published in various international mathematical journals, acquiring great recognition from a broad community of mathematicians worldwide.

Derived from an incredible record as a reviewer for several international mathematics journals, Dr Francisco Bulnes received many invitations to belong to editorial boards as an associate editor or editorial board member.

His incursion into science, the problems of theoretical physics, nanotechnology and electronics has given him worldwide recognition in these areas, where he uses much of his solid training as a mathematician. For example, he holds important positions in specialised nanotechnology and electronics engineering journals (see Figure 66).

Figure 66. Dr Francisco Bulnes as distinguished Honorary Editor of the *World Journal of Electrical and Electronics Engineering* [99].

And, of course, in all areas of mathematics, he is an authority as a universal mathematician (an appellation given to mathematicians who are knowledgeable in all areas of mathematics and current and frontier topics in mathematics). Few mathematicians have this recognition.

Figure 67. Post-Doctoral seminar directed by Dr Francisco Bulnes on spectral curvature and its spectrums. This seminar was celebrated for ten years in Mexico, Ukraine and Russia. The first edition of the book was published in Europe [100].

He leads several maths seminars worldwide, including science seminars with his research groups in Mexico.

Figure 68. International book on complex analysis and applications, including a compilation of chapters about expert mathematicians and topologists. Dr Francisco Bulnes is the editor of this book, with Dr Olga Hachay from Russia, as his co-editor. The book has the international registration numbers ISBN 978-1-83968-360-2, eISBN 978-1-83968-361-9 [101].

The last seminar included works of famous specialist mathematicians in complex variables (see Figure 68).

Dr Bulnes edited a book on topology, an area of mathematics that establishes relations and transformations between spaces with a certain structure depending on their position and considers the structure of the ambient space where these relations exist. This book discusses the various concepts and theories of topology, including diffeomorphisms, immersions, Hausdorff spaces, cobordisms, homotopy theory, symplectic manifolds, the topology of quantum field theory, algebraic varieties, dimension theory, Koszul complexes, continuum theory, and metrizability, among others [102].

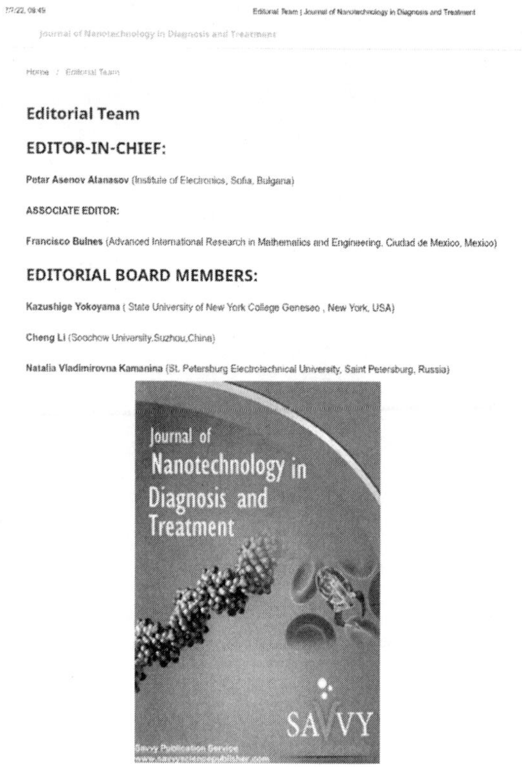
Editorial Team | Journal of Nanotechnology in Diagnosis and Treatment

Home / Editorial Team

Editorial Team

EDITOR-IN-CHIEF:

Petar Asenov Atanasov (Institute of Electronics, Sofia, Bulgaria)

ASSOCIATE EDITOR:

Francisco Bulnes (Advanced International Research in Mathematics and Engineering, Ciudad de Mexico, Mexico)

EDITORIAL BOARD MEMBERS:

Kazushige Yokoyama (State University of New York College Geneseo , New York, USA)

Cheng Li (Soochow University,Suzhou,China)

Natalia Vladimirovna Kamanina (St. Petersburg Electrotechnical University, Saint Petersburg, Russia)

Figure 69. Associate editor position in a *Journal of Nanotechnology* [103].

Figure 70. Dr Bulnes (in the middle), with his beloved systems engineering students from TESCHA, appeared at an event where he was director, principal chairman and first adviser of the Appliedmath III conference [104].

His advantage as a mathematician allows him to create and design essential concepts within other areas of science, such as medicine (see Figure 69), engineering and, as we have said, education as an object of study.

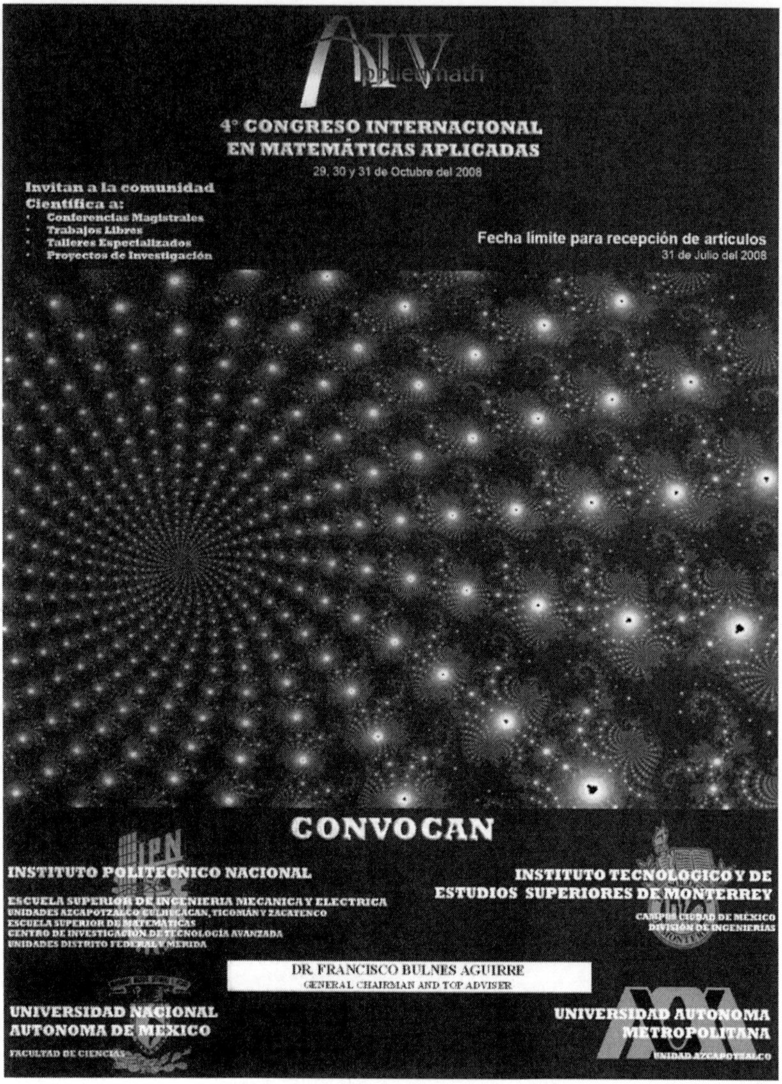

Figure 71. The international event in applied mathematics (Appliedmath), where Dr Francisco Bulnes was director, principal adviser and chair for four years (2005–2008). Many books with ISBN are published on his lectures, talks and reports. The four important universities financially support him [105].

He directed the most important congress of applied mathematics for four years, Appliedmath, attended by many professors and students from many universities in Mexico and abroad (see Figure 70 and Figure 71).

He conducted various nanotechnology, nanomedicine, materials science, electrical and electronic engineering, and quantum mechanics investigations and established new higher education and research methodologies. Thus, he created new paths for research, and everything is framed in a research programme entitled "Research Programme in Mathematical Electrodynamics," where he established electrodynamics, both classical and quantum, as a theoretical-applied basis. The fundamental part is always mathematics. The programme is very ambitious, as it seeks, for example, to create permanent energy transduction devices and electromagnetic propulsion for vehicles and develop a device to cure all diseases under the energy principle.

Dr Bulnes has been searching for and developing these technologies for some time, obtaining partial results and involving young people in his research[1] since he believes that young people are creative and full of imagination, which is lost with age.

The next paragraph of this chapter is by the mathematician Dr Yuri Stropovsvky, a colleague of Dr Bulnes.

5.2. Dr Bulnes' theorems in mathematics in geometry, topology, functional analysis and global analysis

5.2.1. Theorem on unique geometrical pictures in the field theory of different cohomological classes of the sheaves of a derived category on surface

I am Dr Yuri Stropovsvky-Golinsch, a mathematician and close collaborator with Dr Francisco Bulnes. I will give a very brief overview of the aspects of research in pure mathematics developed by Dr Bulnes, trying to be as specific as possible since studying the work of Dr Bulnes covers geometry, algebra and mathematical analysis, which would lead to several volumes as extensive as this book.

In the context of certain derived categories are established correspondences between the moduli stacks and solution classes represented cohomologically under the study of the kernels of differential operators in their classification of the corresponding field equations. Studying the traces

[1] This research and results can be found in different science platforms such as Scopus, Thompson Reuters, Copernicus, CorssRef, Google Scholars, and more.

of these differential operators and classifying them in two groups conformally, we have the corresponding *D*-modules viewed as sheaves of conformal blocks (or co-invariants) (images under a version of the Penrose transform) naturally arising in the framework of conformal field theory. Inside the geometrical Langlands correspondence and in their cohomological context of strings, Francisco Bulnes established a space-time framework through the different versions of the Penrose transforms and their relationship by intertwining operators (integral transforms that are isomorphisms between cohomological spaces of orbital spaces of space-time). Incidentally, Dr Bulnes has a monograph on orbital spaces of Lie groups, their corresponding algebras and their integration on them.

Dr Bulnes obtained the functors that give equivalences of the corresponding categories [106, 107].

The different cycles and co-cycles that he obtained were from the Penrose transform apparatus and their relation with different classes of the Hecke category $H_{G,\lambda}$, causing him to conjecture that the unique geometrical pictures in field theory to different cohomological classes of the sheaves in the derived category $D^{\times}(Bund_G(\Sigma))$, are geometrical objects belonging to the global Langlands category (let monodromic or not) corresponding to a system $Loc_{^LG}(E^{\times})$. The objects included in a category that involves a category of quasi-coherent sheaves on $\boldsymbol{D}G$, of certain fibres on the generalised flag manifolds are λ −twisted D −modules of the flag variety $G/\mathbf{B}$. [108]

Dr Bulnes obtained the functors that give equivalences of the corresponding categories using the geometrical analysis philosophy, inverse problems on geometry and global analysis.

Likewise, the integral geometry methods are applied to deformed categories to obtain correspondences in the geometrical Langlands programme and construct the due equivalences between geometrical objects of the moduli stacks and algebraic objects of the corresponding categories and their ^LG – opers, characterising the solution classes to field theory equations in the belonging cohomological context. Likewise, we have $H^0(g[[z]], \mathbf{V}_{critical}) = \mathbf{C}\left[\mathrm{Op}_{^LG}(D^{\times})\right],$, which is natural in the framework of the integral transforms for generalising the Zuckerman functors. Then Dr Bulnes obtains from different factors the universal functor of derived sheaves of Harish-Chandra, the Langlands geometrical programme in mirror symmetry. This programme is developed around the connections, extensions, and corresponding bundles of geometrical objects.

However, the cosmological problem is to reduce the number of field equations that are resoluble under the same gauge field (the *Verma modules*) and to extend the gauge solutions to other fields using the topological group symmetries that define their interactions [109].

Likewise, Dr Bulnes affirms that the extension can be given by a global Langlands correspondence between the Hecke sheaves category $H_{\hat{G},\infty}$on an adequate moduli stack and the holomorphic $^{L}G-$, bundles category with a special connection (the *Deligne connection*). The corresponding $D-$modules may be viewed as sheaves of conformal blocks (or co-invariants) (images under a generalised version of the Penrose transform) naturally arising in the framework of conformal field theory. Using extensions of Verma modules, Dr Bulnes obtained the following results and a classification table:

Theorem 5. 2.1. 1 (F. Bulnes). The derived category of quasi $-G-$equivariant $--D_{G/H}-$modules shaped with the generalised and extended Verma modules given by $^{L}\Phi^{\mu}(M) = M \boxtimes \boldsymbol{\rho}^{\mu}(\mathbf{V}),\ \forall \mathbf{V} \in (\boldsymbol{Loc}_{L})$, can be identified as a critically twisted sheaf category of $D-$modules on the moduli stack $\mathrm{Bun}_{G,y,}\forall y \in X$ (*singularity*) identified by the Hecke category $H_{G,K,y}$, (the geometrical Langlands correspondence) if this is an image of integral transforms acting on ramifications of the Hecke category $H_{G}, \forall \boldsymbol{\lambda} \in \boldsymbol{h}^{*}$ (for example$H_{G,\lambda}$) on the flag variety $G/\mathbf{B}$,, with weight corresponding to twisted differential operators on $\mathrm{Bun}_{G,y.}$

Proof. [109].

The application of the Penrose transform on $^{L}\,\mathrm{Bun}_{G}$, which are generalised Verma modules of the Langlands geometrical equivalences $D_{BRST}\left(\mathrm{Oper}_{^{L}G}^{\leq n}(D_{y})\right) \cong D(\mathrm{Bun}_{G}(\Sigma))$, demonstrated in [109], is the theorem on the Hecke classes as solutions.

By conjecture 8, given in [106], the unique objects that are coefficients of the cohomological space of dimension 0, that is to say, spaces $\left(H^{0}(X, O)\right)$, are the Verma modules $\mathbf{V}_{\lambda}(\chi), \forall \chi \in Op_{^{L}G}^{\lambda}$.

But $\mathrm{D}^{b}(\tilde{g}_{K_{C}} - \mathrm{mod}_{\chi}) \cong D^{b}(Q\mathrm{Coh}(\mathrm{Sp}_{u}^{DG}))^{2}$ then the Hecke category is of the Verma modules $^{L}\Phi^{\mu}\left(\mathbf{V}_{w(\lambda+\rho)-\rho}(\chi_{y})\right), \forall w \in W$, which are the ramifications of the Hecke category $H_{G}, \forall \lambda \in h^{*}$.

To know the cycles and co-cycles explicitly and under the geometrical arguments of their Penrose transforms, Dr Bulnes gives the following classification table for consulting:

---

[2] Derived category of quasicoherent sheaves on a suitable DG.

Derived Twisted Category $H_{G,\lambda}$	**Geometrical Image: Holomorphic** ^{L}G **−bundles**	**Co-cycles**
$H_{G,\lambda}^{aff}$	$\mathbf{C}^{\times}$ −Bundle	$H\backslash\tilde{G}/K$ Complex Hyper-lines
$H_G \cong (B\backslash G/B)$	$D^{\times}$ −Bundle	$B\backslash\tilde{G}/B$ Deformable Cycles or Orbifolds
$H_G^{mon} \cong (N\backslash G/N), \lambda = 0$	$\tilde{D}^{\times}$ −Bundle	$N\backslash\tilde{G}/N$ Hyper-planes or hypersurfaces
$H \cong (G(F)\backslash G(\mathbf{A}_F)/G(O))$	D_{BRST} −Bundle	$G(F)\backslash G(\mathbf{A}_F)/G(O)$ Strings

Table 1. Some Hecke categories and their ramifications.

After we generalise the functors $R\Gamma_K^H$, to the functors $^{L}\Phi_{\boxtimes}^{\mu}$, we obtain certain correspondences between cycles and co-cycles on derived categories of corresponding $D_{\mathbf{P}}$ −modules that are quasi-coherent D_{λ} −modules (with the corresponding character of the Hecke category [106-109]) whose deformed image corresponds to some ramifications [109, 110], which can be classified by the following theorem given in [108, 109]:

Theorem 5. 2. 1. 2 (F. Bulnes). Consider the classes of the Hecke category $H_{\hat{G},\infty}$, satisfying the geometrical Langlands correspondences given by

$$D_{BRST}\left(\mathrm{Oper}_{^{L}G}^{\leq n}(D_y)\right) \cong D\big(\mathrm{Bun}_G(\Sigma)\big), \qquad (13)$$

(as was established in [2]). The unique geometrical pictures in field theory to different cohomological classes of the sheaves in $D^{\times}\big(\mathrm{Bun}_G(\Sigma)\big)$, , are the co-cycles (images under different Penrose transform versions):

a) Orbifolds or,
b) Strings as twisted hyper-lines and twisted hyper-planes or,
c) Super-twistor surfaces (from complex 2- and 3-dimension spaces).

Proof. [108].

Due to the importance of the results obtained in algebraic geometry and mathematical physics on derived categories in the Langlands programme, one of his works was translated into Russian, as a great recognition (see Figure 72).

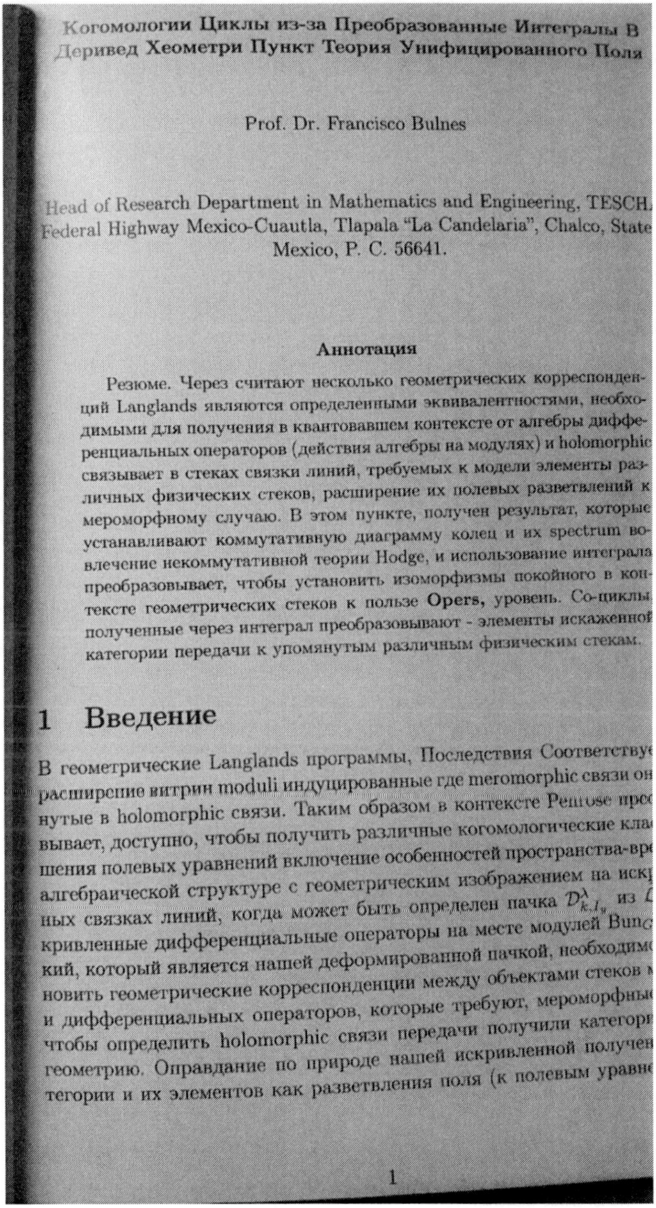

Когомологии Циклы из-за Преобразованные Интегралы В Деривед Хеометри Пункт Теория Унифицированного Поля

Prof. Dr. Francisco Bulnes

Head of Research Department in Mathematics and Engineering, TESCH, Federal Highway Mexico-Cuautla, Tlapala "La Candelaria", Chalco, State Mexico, P. C. 56641.

Аннотация

Резюме. Через считают несколько геометрических корреспонденций Langlands являются определенными эквивалентностями, необходимыми для получения в квантовавшем контексте от алгебры дифференциальных операторов (действия алгебры на модулях) и holomorphic связывает в стеках связки линий, требуемых к модели элементы различных физических стеков, расширение их полевых разветвлений к мероморфному случаю. В этом пункте, получен результат, которые устанавливают коммутативную диаграмму колец и их spectrum вовлечение некоммутативной теории Hodge, и использование интеграла преобразовывает, чтобы установить изоморфизмы покойного в контексте геометрических стеков к пользе **Opers**, уровень. Со-циклы, полученные через интеграл преобразовывают - элементы искаженной категории передачи к упомянутым различным физическим стекам.

1 Введение

В геометрические Langlands программы, Последствия Соответству расширение витрин moduli индуцированные где meromorphic связи он нутые в holomorphic связи. Таким образом в контексте Penrose прео вывает, доступно, чтобы получить различные когомологические кла шения полевых уравнений включение особенностей пространства-вре алгебраической структуре с геометрическим изображением на иск ных связках линий, когда может быть определен пачка $\mathcal{D}^{\lambda}_{k,I_y}$ из кривленные дифференциальные операторы на месте модулей Bun кий, который является нашей деформированной пачкой, необходим новить геометрические корреспонденции между объектами стеков и дифференциальных операторов, которые требуют, мероморфны чтобы определить holomorphic связи передачи получили категор геометрию. Оправдание по природе нашей искривленной получен тегории и их элементов как разветвления поля (к полевым уравн

1

Figure 72. Dr Bulnes's research on derived categories translated into Russian idioms [38].

5.2.2. Searching to Generalise the Curvature Concept as Tempered Distribution or a Generalised function

This started with the study of curvature as a geometrical invariant and field observable in gravitation in the classical studies of differential geometry and gravitation theory, including Newton's general gravitation law and Einstein's general relativity. Curvature is conceptualised as measuring the variable rate, velocity or the change of direction of a point in a curve, surface or space. On the other hand, in gravitation, the curvature is a perception of gravity in the presence of matter; that is to say, it is the geometrical scenario created in the space obtained for the presence of matter which generates a gravity field. Curvature is the observed result of gravity.

In 2001, the results on curvature were published, searching for a tempered distribution or generalised function to study the universe modelled as a real or complex Riemannian manifold without singularities. The basic idea was to obtain the curvature value for certain integrals on a light geodesic, where these integrals are given through a functional transform called the Radon transform, which realises the tomographic actions of a space. This study is considered the 2-form of curvature and is defined as the sectional curvature. The curvature stays as the image of a smooth embedding of a section (a cycle of the manifold of space in question) to be the co-cycle in the space of tempered distributions (the curvature function of this cycle). The result was the following:

Theorem (F. Bulnes, UNAM, 2001) 5. 2. 2. 1. Let be $\sigma: \Sigma \to D'(M)$, a smooth embedding of a smooth finite dimensional manifold Σ, in the distribution space $D'(M)$, and let be $R_\sigma: C_c^\infty(M) \to C^\infty(\Sigma)$, the Radon transform associated with this embedding. Then the extrinsic curvature of $\sigma(\Sigma)$, in $D'(M)$, is the Hessian composed with the generalised Radon transform.

Subsequently, several corollaries and applications in electronics, systems and signal analysis and the field theory of the concept of curvature [19-24] are obtained to determine the concept of energy of curvature (Figure 73).

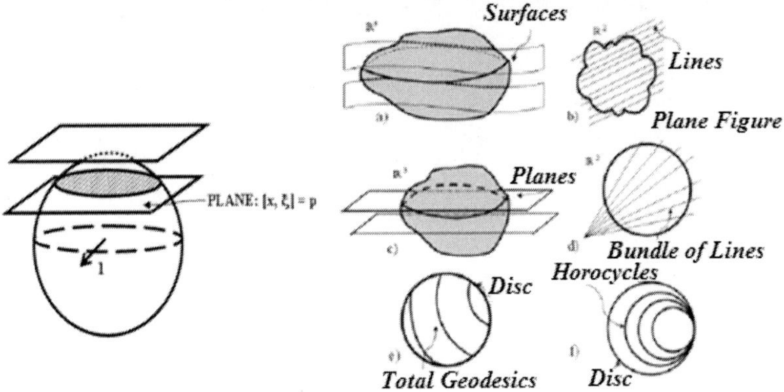

Figure 73. As mentioned, the generalisation of curvature obtained through theorem 2.1 determines the curvature of a space or body through cycles or parts of a space, where the space curvature or body curvature is the integration of all these cycles. The curvature obtained is in tempered distribution space. Likewise, the inverse Radon transform is applied for all the co-cycles of curvature obtained, which is its curvature. Figure 73 shows the topography through different space or body cycles, determining the curvature value.

A recompilation of universe research through field theory is in the special and luxurious edition of a book published by Cambridge Scholars Publishing in the United Kingdom (see Figure 74). This book contains recent and important results on the deep study of the universe through one of the field observables that will have a great impact in the next 100 years on space-time research and the creation of future technologies. It will appeal to mathematicians, natural scientists, cosmologists, theoretical physicians, particle physicians, electronics researchers and postgraduate students in mathematical physics dedicated to studying the universe. ISBN: 1-5275-8728-2, ISBN13: 978-1-5275-8728-1 [113].

Figure 74. Treatise and monograph on universe studies from the perspective of field torsion, realised by Dr Francisco Bulnes [113].

These investigations evolved to create a "Field Torsion and its Applications" research programme, where Dr Bulnes involves torsion as a second curvature. He also developed several conjectures about the universe and the creation of technology based on principles of field torsion. For example, some results and conjectures have been the following [26]:

Theorem. 5. 2. 2. 2 (F. Bulnes) We consider the spectral curvature $\kappa(\omega_1, \omega_2)$, in a curved surface $\Omega \subseteq \mathbf{R}^3$. The functional torsion due to the force $\delta(t_1, t_2)$, that moves a particle along the height $2\pi b$, and helix of angle $\cot\theta$, is given by:

$$\tau(t_1, t_2) = \pm 2\pi \cot\Psi F\{K(t_1,\ t_2)\}\delta(t_1,\ t_2), \qquad (14)$$

and some corollaries as:

Corollary (F. Bulnes) 5. 2. 2. 1. The torsion energy is its curvature energy.

Or some conjectures as:

Conjecture 5. 2. 2. 1. The curvature in the spinor-twistor framework can be perceived with the appearance of the torsion and the anti-self-dual fields [27].

Conjecture 5. 2. 2. 2 (F. Bulnes). The torsion is the geometrical invariant of the interaction between energy and space.

We would need several volumes like this book to analyse all of them. Dr Bulnes has researched the concept of curvature energy to create sensors of various types and for different applications (Figure 75).

Proposition 5. 2. 2. 1. The Radon Transform of the Gaussian curvature of light signals on the polar disk is their mean curvature.

Figure 75. UK-Promotional material for the international research programme on field torsion [114].

In the deep study of the universe [113], we need to answer the following questions:
Why is torsion required to study the mechanisms of matter/anti-matter in gravity and neutrinos/anti-neutrinos in matter production to obtain gravity? What species of neutrinos are relevant in this work?

To contemplate the following summary on torsion connections, it is necessary to produce a Dirac model that explains the gravity with torsion from the neutrinos and other fermions. First, we can consider the connection given in [31, 113] with $J^{\mu}_{(A)} = \overline{\psi}\gamma_5\gamma^{\mu}\psi$. In the case $N = 1\ SUGRA$, when the connection is

$$\omega_{\mu}^{ab} = \omega_{\mu}^{ab}(e, \psi) = \omega_{\mu}^{ab}(e) + \kappa_{\mu}^{ab},$$

where

$$\kappa_{\mu\alpha\beta}(\psi) = \tfrac{1}{4}\left[\overline{\psi}_{\alpha}\gamma_{\mu}\psi_{\beta} + \overline{\psi}_{\mu}\gamma_{\alpha}\psi_{\beta} - \overline{\psi}_{\mu}\gamma_{\beta}\psi_{\alpha}\right],$$

and the spinor axial current is

$$J^{\mu} \equiv \frac{1}{2}e^{\mu\nu\alpha\beta}\overline{\psi}_{\nu}\gamma_{\alpha}\psi_{\beta},$$

and in the case $N = 2$ $SUGRA$, the connection is the same as in ω_{μ}^{ab}, with the difference that $\kappa_{\mu\alpha\beta}(\psi)$, is given for

$$\kappa_{\mu\alpha\beta}(\psi) = \tfrac{1}{4}\left[\overline{\psi}^{I}_{\nu}\gamma_{\mu}\psi_{I\beta} + \overline{\psi}^{I}_{\mu}\gamma_{\alpha}\psi_{I\beta} - \overline{\psi}^{I}_{\mu}\gamma_{\beta}\psi_{I\alpha}\right],$$

with the spinor axial current:

$$J^{\mu} \equiv e^{\mu\nu\alpha\beta}\overline{\psi}^{I}_{\nu}\gamma_{\alpha}\psi_{I\beta},$$

in the case of $N = 14$ $SUGRA$, the connection is:

$$\omega_{\mu ab} = \omega_{\mu ab}(e, \psi, \Lambda) = \omega_{\mu ab}(e) + \kappa_{\mu ab}(\psi),$$

where

$$\kappa_{\mu\alpha\beta} = \kappa_{\mu\alpha\beta}(\psi) + \kappa_{\mu\alpha\beta}(\Lambda),$$

The term given by $\kappa_{\mu\alpha\beta}$, is the same as

$$\kappa_{\mu\alpha\beta}(\psi) = \frac{1}{4}\left[\overline{\psi}^{I}_{\nu}\gamma_{\mu}\psi_{I\beta} + \overline{\psi}^{I}_{\mu}\gamma_{\alpha}\psi_{I\beta} - \overline{\psi}^{I}_{\mu}\gamma_{\beta}\psi_{I\alpha}\right],$$

and the term $\kappa_{\mu\alpha\beta}(\Lambda)$, is given as:

$$\kappa_{\mu\alpha\beta}(\Lambda) = -\frac{1}{4e}\in_{\mu\alpha\beta\sigma}\overline{\Lambda}_{I}\gamma^{\sigma}\Lambda^{I},$$

Here $J^{\mu}(\Lambda) = e\overline{\Lambda}_{I}\gamma^{\mu}\Lambda^{I}$.

There are some ambiguities in the Einstein-Cartan theory with the Immirzi parameter. For example, total derivatives can modify actions without effects on-shell physics (equations of motion). But such terms may play a role in quantum physics. On the other hand, in pure gravity without torsion, one can add the following identically zero term due to Bianchi identities of symmetry properties of the Riemann tensor:

$$\epsilon^{\mu\nu\rho\sigma}\, R_{\mu\nu\rho\sigma}(\varpi) = 0,$$

no longer true in the presence of torsion. For example, the action

$$S_{Holst} = -\frac{\beta}{4\kappa}\int d^4x e e_a^{\mu} e_b^{\mu}\, \epsilon_{cd}^{ab}\, R_{\mu\nu}^{cd},$$

where the tensor $\tilde{R}_{\mu\nu}^{ab} = \epsilon_{cd}^{ab}\, R_{\mu\nu}^{cd}$, is the dual of the curvature tensor, is nontrivial in torsionful geometries.

But in the torsion case, the movement equations [113] imply:

$$T^{\mu} = \frac{3\kappa}{4}\frac{\beta}{\beta^2+1}J_{(A)}^{\mu},$$

$$S^{\mu} = \frac{3\kappa}{\beta^2+1}J_{(A)}^{\mu},$$

$$q_{\mu\nu\rho} = 0,$$

where it is clearly inconsistent due to the equality between a vector T^{μ}, with the pseudo-vector $J_{(A)}^{\mu}$.

The Holst term needs modification by adding fermions to become a total derivative [31, 113].

The fermion piece (*the fermion key*) that will be added to the Holst action is:

$$S_{Holst-fermion} = \frac{\alpha}{2}\int d^4x e\left(\overline{\psi}\gamma^{\mu}\gamma_5 D_{\mu}(\omega)\psi + \overline{D_{\mu}(\omega)}\gamma^{\mu}\gamma_5\psi\right),\quad \alpha = cont,$$

together with Dirac kinetic term, the fermion action then reads:

$$S_{Dirac-Holst-fermion} = \frac{i}{2}\int d^4x e\left(\overline{\psi}\gamma^{\mu}(1-i\alpha\gamma_5)D_{\mu}(\omega)\psi + \overline{D_{\mu}(\omega)}\gamma^{\mu}(1 - i\alpha\gamma_5\psi\right),$$

If $\alpha = \beta$, the *Holst total gravity-fermion* term becomes total derivative, that is to say:

$$S_{Holst-total} = -\frac{\beta}{2}\int d^4x\left[I_{NY} + \partial_\mu J^\mu_{(A)}\right],$$

which establishes an equivalent to the Einstein-Cartan theory independent of the Immirzi parameter β. In $S_{Holst-total}$, we consider I_{NY}, as the *Nieh-Yan topological invariant density* which creates the unique Lorentz invariant torsion structure, which has the form:

$$I_{NY} \equiv e^{\mu\nu\rho\sigma}\left(T^a_{\mu\nu}T_{\rho\sigma a} - \frac{1}{2}e^a_\mu e^b_\nu R_{\rho\sigma ab}(\omega)\right),$$

which, in simplified form, is written as

$$I_{NY} \equiv e^{\mu\nu\rho\sigma}\partial_\mu T_{\mu\nu\sigma}.$$

5.2.3. Some research on Motivic Cohomology[3]

Francisco Bulnes' study, following his research programme on derived categories and integral geometry in the study of the universe (where the latter is modelled as a complex Riemannian manifold including singularities as holes, knots and cusps), was focused on the derived tensor product whose functors have images as cohomology groups that are representations of integrals of sheaves represented for their pre-sheaves in an order modulo k. This study is remounted to the $K-$ theory on the sheaves cohomologies constructed through pre-sheaves defined by the tensor product on commutative rings. This study intends to establish a methodology through commutative rings and their construction of a total tensor product $\otimes^{\mathbf{L}}$,[4] on the category $\mathrm{PST}(k)$, considering extensions of the tensor products $\otimes_{R(A)}$, to obtain a resolution in the projective sense of infinite sequences of modules of Étale sheaves. These sheaves are pre-sheaves of Abelian groups in the category of smooth-separated schemes restricted to the scheme X.

Likewise, Dr Bulnes applies the derived tensor products to determine the tensor triangulated category $\mathrm{DM}^{\sim}_{\acute{e}t}(k, \mathbf{Z}/m)$, of Étale motives

[3] Postdoctoral Research supported by State of Mexico Council of Scientific Research COMECYT-077/111/21.

[4] **L**, is a Lefschetz motive $\mathbf{Z}(1)$, [116].

to be equivalent to the derived category of discrete $\mathbf{Z}/m$ −modules over the Galois group $G = Gal(k_{sep}, k)$, which determines the equivalence of functors of tensor triangulated categories[5] [115].

Then the mean result of derived tensor products will be in the tensor triangulated category $DM_{Nis}^{eff-}(k, R)$, of effective motives and its subcategory of effective geometric motives, $\mathrm{DM}_{gm}^{eff-}(k, R)$. Likewise, the motive $M(X)$, of the scheme, X, is an object of $\mathrm{DM}_{Nis}^{eff-}(k, R)$, and belongs to $\mathrm{DM}_{gm}^{eff-}(k, R)$, if X, is smooth. However, this requires using cohomological properties of sheaves associated with homotopy invariant pre-sheaves with transfers for Zariski topology, Nisnevich and cdh topologies.

Finally, all this treatment develops motivic cohomology to establish a resolution in the field theory to solutions to the field equations described in a general form (using complex and real differential forms)

$$dda = 0, \qquad (15)$$

incorporating singularities in the complex Riemannian manifolds where singularities can be studied with deformation theory through operads, motives and deformation quantisation. This is precisely the following theorem given by Dr Bulnes:

Theorem (F. Bulnes) 5. 2. 3. 1. Suppose that **M**, complex Riemannian manifold with singularities. Let be X, and Y, smooth projective varieties in **M**[6]. We know that solutions of the field equations $dda = 0$, are given in a category $\mathrm{Spec}(\mathrm{Sm}_k)$. Solutions in the context of the quantum field equations for $dda = 0$, are defined in a hyper-cohomology on $\mathbf{Q}$ −coefficients from the category Sm_k, defined on a numerical field k, considering the derived tensor product $\otimes_{\acute{e}t}^{tr}$, of pre-sheaves. Then the following triangulated tensor category scheme is true and commutative [115]:

---

[5] **Theorem.** If $1/m \in k$, $(L \otimes_L)$, is a tensor triangulated category and the functors

$$\mathrm{D}^-(G, \mathbf{Z}/m) \overset{\pi*}{\to} L \to \mathrm{D}(W_A^{-1}) = DM_{\acute{e}t}^{eff,-}(k, \mathbf{Z}/m),$$

are equivalences of tensor triangulated categories. Here $\pi *$, is a triangulated functor from $\mathrm{D}^-(G, \mathbf{Z}/m)$, until the category $D^-(Sh_{\acute{e}t}(Cor_k, \mathbf{Z}/m))$.

[6] Singular projective varieties useful in the quantization process of the complex Riemannian manifold. the quantization condition compact quantizable Kaehler manifolds can be embedded into projective space.

$$\begin{array}{ccc} & DQFT & \\ i\swarrow & & \searrow F \\ \mathrm{MD_{gm}}(\mathbf{Q}) & \leftrightarrow & MD(O_Y) \end{array}, \qquad (16)$$

Dr Bulnes affirms that just the category $DQFT$, can be defined as the motives in a hypercohomology from the category Sm_k, defined as:

$$Hom_{\mathrm{DM_{gm}^{eff}}(k)}(m(X), \mathbf{Q}(q)[p]) \cong H_{Nis}^{*}(X, \mathbf{Q}(q)) = \mathbf{H}^{p,q}(X_{Nis}, \mathbf{Q}(q)), (17)$$

which comes from the hypercohomology

$$H_L^{p,q}(X, \mathbf{Q}) = \mathbf{H}^{p,q}(X, \mathbf{Q}), \qquad (18)$$

Dr Bulnes observes that if a Zariski sheaf of $\mathbf{Q}$ −modules with transfers F, is such that $F = H^q C$, for all C, a complex defined on $\mathbf{Q}(q)$ −modules, being a special case when $C = \mathbf{Q}(q)$, where the cohomology groups of the isomorphism $H_{\acute{e}t}^{p}(X, F_{\acute{e}t}) \cong H_{Nis}^{p}(X, F_{Nis})$,, can be vanished for $p > \dim(Y)$; then survive a hypercohomology $\mathbb{H}^q(X, \mathbb{Q})$. Finally, considering $\mathrm{Spec}(\mathrm{Sm}_k)$, Dr Bulnes establishes the quantum version of this hyper-cohomology with additional work on moduli stacks of the category $\mathrm{M}od_B$.

Dr Bulnes obtained another interesting result from this research in the following theorem.

Theorem (F. Bulnes) 5. 2. 3. 2. $H^*(GL(n,k))$, has the decomposing components $H^i(X)$, that are hyper-cohomology groups corresponding to solutions as $\mathbf{H}$ −states in $\mathrm{Vec}_{\mathbb{C}}$, for field equations $dda = 0$.

The details of the demonstration can be found in [117].

5.3. International Seminars Directed by Dr Bulnes, and the Frontiers of Knowledge

Dr Stropovsvky here. I want to say that I have seen many aspects of his leadership and guidance for other colleagues of his since I have collaborated with him on various projects and seminars. From around the world, many young scientists and recent PhD graduates communicate daily with Dr Bulnes, asking for diverse support and help [118, 119]. This support and help can be for an opinion on some academic work to financial aid for publishing an article. Other recent PhD graduates in mathematics, physics or engineering invite him to appear as a co-author in their publications, adding weight to their submission of a paper to a scientific journal. However, his professional ethics have not allowed him to accept as he cannot agree to co-author a paper he did not work on. This is worth noting because in the academic world, especially in many developing countries, there is the vice of co-authoring material one has not worked on. This is a common vice among academics in Latin America, where it is intended to earn stipends and scholarships through academic production. For example, the Conacyt in Mexico or the CONICET in Argentina.

Dr Bulnes has helped resolve scholarly controversies featured in papers, including some maths monographs. Dr Bulnes has had to intervene as a mediator and arbitrator in these discussions and controversies, honouring his investiture as Ambassador of Peace, a decoration bestowed in Germany by ODAEE and the Red Cross in 2018.

Some young mathematics doctors who have sought and had support from Dr Bulnes have been from countries such as Russia, Ukraine, the Czech Republic, Iran, Pakistan, India, China and Singapore, and in some cases, have even requested very specific advice on specialised maths topics, in the US and UK.

This led to the creation of several international post-doctoral research seminars on topics such as curvature and programmes on derived categories in the so-called Langlands geometric programme. Of all this, Dr Bulnes has published papers generated by some attendees at these seminars. These have lasted one to three years on average, with results published in maths journals, books and monographs, including some book chapters endorsed by Dr Bulnes for special compilations of advanced maths topics.

One such seminar is the Mexico-Ukraine-Russia seminar, which has been celebrated for four years and continues today.

Another seminar of mathematicians in Iran proposed to relate the CW-complexes used in his algebraic geometry programme (see Figure 76) with the Cohen complexes, MacCauley modules and Stanley conjecture [120, 121].

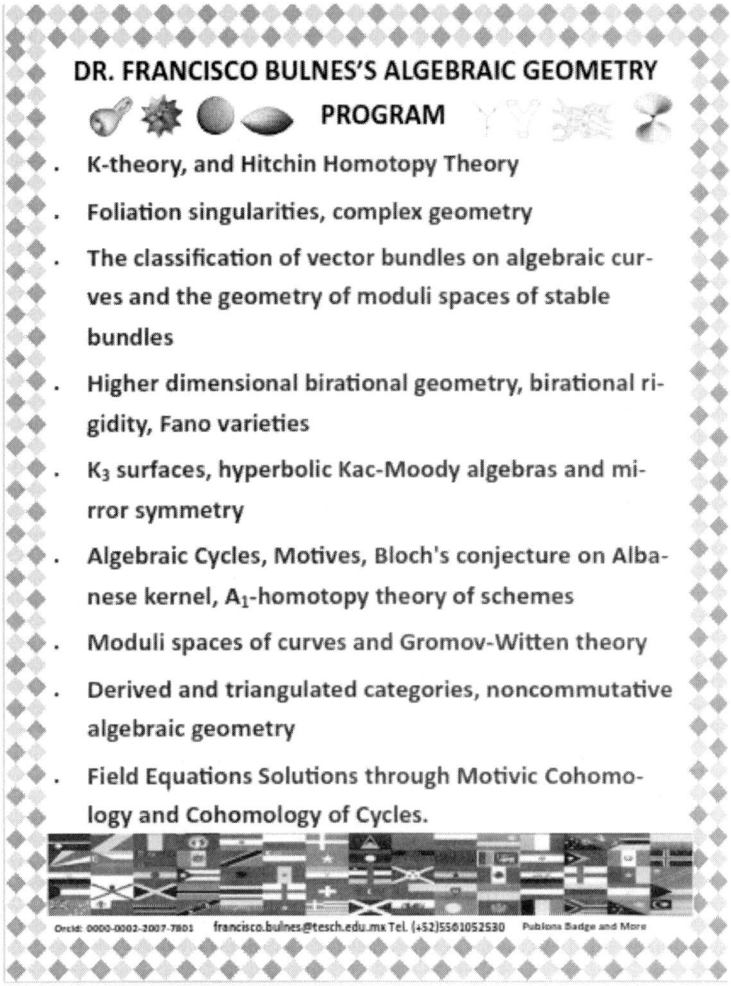

Figure 76. Research programme in algebraic geometry by Dr Francisco Bulnes [121].

Another mathematics programme has been nanotechnology and mathematical research theory, where the latter has been developed as a diploma course of 250 in six modules taught at TESCHA. This is open to anyone who wants to learn to do serious and internationally renowned research.

In his post-doctoral activities, Dr Bulnes directed and advised on a special issue in the Langlands Programme [122] (see Figure 77 and Figure 78).

Figure 77. Post-Doctoral seminar celebrated in Russia and Ukraine entitled "Integral Geometry Methods on Derived Categories in the Geometrical Langlands Programme" and published in a special issue of the *Pure and Applied Mathematics Journal*, USA, in 2014.

Figure 78. Results obtained during the seminar and published in a special issue of *Pure and Applied Mathematics Journal* ISSN: 2326-9790, edited and printed in the US.

5.4. Much of his experience was shared with other colleagues worldwide.

Dr Bulnes has been invited to more than 100 countries (on five continents) to participate as a speaker, lecturer or exhibitor of his theories in mathematics, theoretical physics, nanotechnology and advanced electronics and mechanical engineering, where he has participated in different modalities, either in person, remotely or virtually, using the latest technological tools. He has also been interviewed at important international events, such as in Dubai, UAE[7], where he gave a talk and visited the important universities of UAE at the Male Campus (one of the most expensive universities in UAE[8]). Afterwards, he took a short trip to Kuwait and Oman to discuss collaborating with other scientists. Dr Aziz reproduced and published the interview with Dr Bulnes in Dubai, United Arab Emirates, featured in 2012 on the website of the Bentham Science Newsletter (please see Figure 79) and (see details in [123]).

---

[7] United Arab Emirates.

[8] Many of the young people who study there are children of Sheikhs and sultans from all over the Arabia.

Figure 79. Poster for the scientific newspaper announcement of the interview in Dubai, United Arab Emirates, in 2012. This interview was published online in the *Bentham Science News Letter* (BenSci) [123]. The complete interview can be read in [31, 124].

The research in nanotechnology and nanomedicine left a great impression on Arab scientists.

The vision of Dr Francisco Bulnes goes beyond the matter and molecular levels because, as he says, any illness begins at the atomic level through collateral atomic damage. Voracious protons eat unstable electrons, whose electronic information through photons is erroneous, and provokes the exit of these electrons attracted to voracious protons outside of the atom.

The electronic hypothesis in this respect can be detected for field variations of the vital field or deviations of this field in their geometry. Integral geometry elements are incorporated to explain and determine this field deviation for the energy spectra. Perhaps Dr Bulnes had been thinking about different organic systems using these different spectral resonances in the energy scanner for human bones.

Dr Francisco Bulnes shows through very advanced mathematics the interchange of particles required to spill in the damaged field to correct and restore this field and obtain a cure. He uses a version of the Feynman integrals to explain the interchanges and actions through the energy paths of the particles (see Figure 80).

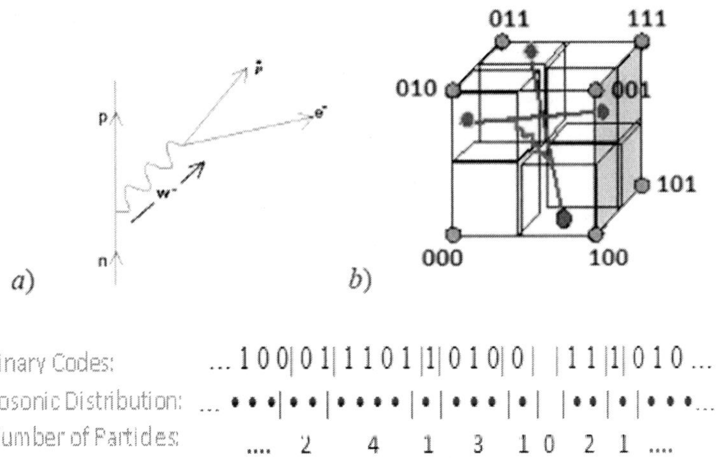

Figure 80. The nanomedicine research of Doctor Francisco Bulnes [125, 126].

However, this research was beyond organic suffering, concluding that all suffering or illnesses originate in the mind-body. Likewise, Dr Bulnes concluded in research presented at an international conference, "…all is reduced to correct and recover The WILL…" and "…the will is all (*is the support of all*)".

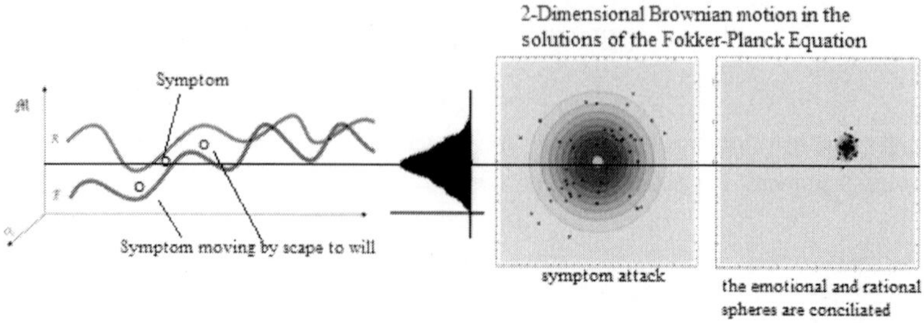

Figure 81. A symptom of a singularity in the space-time of the unlimited quantum region of the mind

A symptom (Figure 81) arises when there is a separation between $\boldsymbol{E}-$ and $\boldsymbol{R}-$curves of the free discerning mind and forms a cumulus of energy or energy bale with a certain weight $W(s,t)$, constructed through $a_i{}^{\alpha_i}$ (*the order of nano-joules*) [127]. This symptom, as energy, produces a disturbance in the space-time of the mind. Likewise, a perturbation 2-dimensional model is shown in Figure 82.

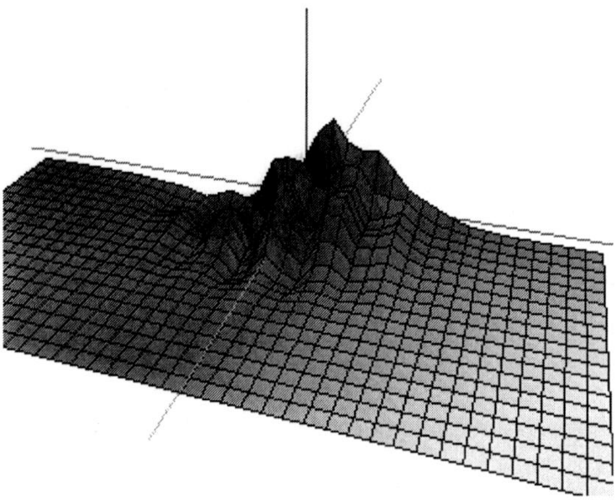

Figure 82. The appearance of the symptom as a non-linear perturbation.

For string theory and quantum electrodynamics (QED), Dr Francisco Bulnes obtained the following relationship to determine the dimension of a symptom:

$$\dim Sym = \prod_i a_i{}^{\alpha_i}, \quad (19)$$

where $a_i^{\alpha_i}$, are factors of energy that give value to the weight $W(s,t)$, which is developed when $s > 0$, with $s \leq t$.

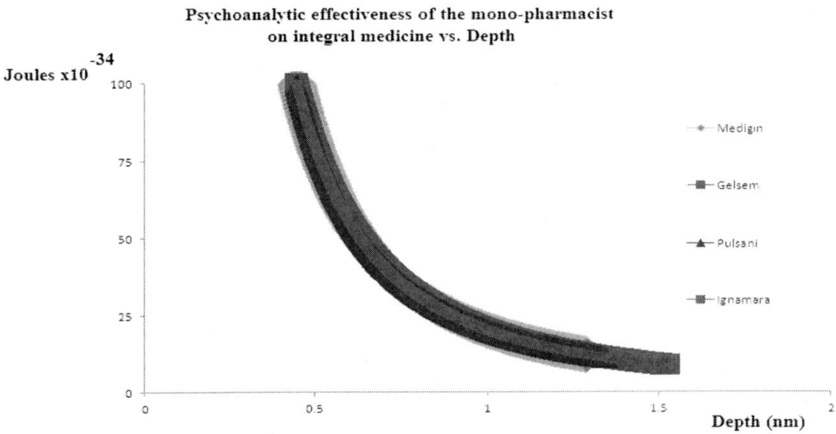

Figure 83. Psychoanalytic effectiveness versus depth.

This energy was calculated with the following experimental equation $\mathbf{M} = \mathbf{R}^a\mathbf{E}^l$, from the quantum integral transform studied in [128] considering the minimum energy of each mono-pharmacist, which is predicted by the *theorem 1*, from [128] (The formula is a short form of the integral transform where $\mathbf{R}^a$, is $\dim\Lambda(\alpha)$, and $\mathbf{E}^l$, is between brackets [128]). Dr Bulnes affirms that anxiety can be displaced using quantum actions $\Im_Q$, through four fundamental quantum mono-pharmacists of the Gelsium group: Gelsem, Ignamara, Medigin and Pulsani (Figure 83). Dr Bulnes explains and determines his theory on the cure and recomposing of the mind through these results. Dr Bulnes characterises a symptom as a singularity in the space-time of the unlimited quantum region of the mind.

Lemma 5. 4. 1 (Bulnes). Let $\mathrm{M} = M \times I_t$, the unlimited space of the quantum space (Fock space [128, 129]). A particle $x(t)$, that is focalised by a bad evolution given for the energy load function $w(s,t)$, is given for

$$x(s) = \int_{\mathbf{H}} \phi(x')x(t)dx(t) = \int_{-\infty}^{\infty} \delta(t-s)x(t)dt, \qquad (20)$$

Then time $t = s$, begins the singularity.
Proof. [129].

If we consider a conscience operator with singularity $O_c(*)$, for the presence of an energy load $w(s,t)$, then the *elimination* of the singularity $*(s)$, is given by the theorem [129]:

Theorem. 5. 4. 1. (F. Bulnes). Let us consider a conscience operator with singularity $O_c(*)$, for the presence of an energy load $w(s,t)$. Then the *elimination* of the singularity $*(s)$, is given for

$$x(t) = correction + restoring = \int_{X(C)} O_c(*(s))w(t,s)dt =$$

$$\dim\Lambda(\alpha) \int_C \left\{ \frac{1}{A} \prod_{j=1}^{\infty} \left\{ \int_{-\infty}^{\infty} \phi(n_j)F(n_j) \right\} dx(t) \right\}, \qquad (21)$$

where $\dim\Lambda(\alpha)$, is the *Neumann dimension* corresponding to the *Weyl camera* of the roots α_j, [129, 130] used in the rotation process to eliminate the deviation [129] created by the singularity.

Proof. [129, 130].

Arguments for algebra and homotopies meeting demonstrate the nature of the sum "*correction* + *restoring*" created by the path integral established in [131]. Then we can enunciate the pair of quantum integral transforms as [129, 132]:

$$Q\{x(t)\} = \int_{X(C)} O_c(x(s))w(t,s)ds = \int_{X(C)} O_c(x(t))d\phi, \qquad (22)$$

$$Q^{-1}\{*(t)\} = dim\Lambda(\alpha) \int_{X(C)} O_c(*(s))w(t,s)dt = \int_{X(C)} O_c(*(s))dx, \quad (23)$$

The field theory and some ideas in mathematical nanotechnology developed by Dr Bulnes are used to create a field torsion theory on fine and advanced aspects in electronics and develop a sensor capable of detecting and measuring, at least indirectly (using inverse problems in mathematics) the field torsion in space-time [26, 133] (see Figure 84).

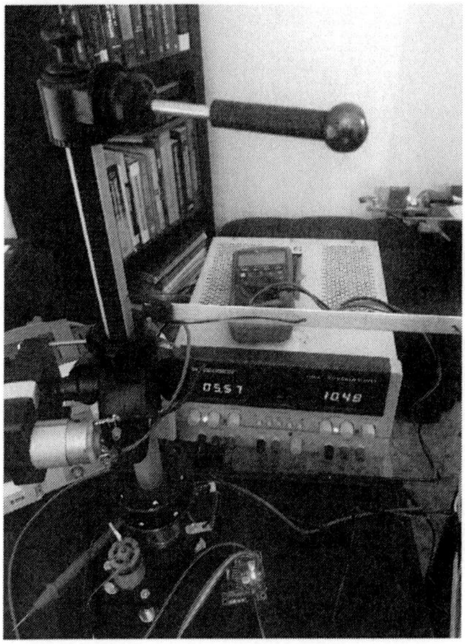

Figure 84. A research group directed and led by Dr Bulnes creates and develops an electronic device on a Hall sensor to measure and detect field torsion, with measures in the [133] space $L^2(L_{EM}(z(t)), \mu)$.

5.5. God's Language

I am Daniel Alvarez, an academician, electronics engineering professor and artificial intelligence specialist. All science matters interest me.

The entertaining Disney film *Donald in the Country of Mathematics* ends sublimely with Galileo Galilei's phrase: ***"Mathematics is the alphabet with which God has written the universe"*** [134]; after, the film explains the usefulness of mathematics in all areas in which human beings develop. In the Book of Genesis, this phrase is attributed to God: ***"And God said 'Let there be light,' and there was light."*** [135]. I remember seeing a poster with the following legend as a child (Figure 85).

...and God says:

$$\vec{\nabla} \cdot \vec{E} = \frac{\rho}{\varepsilon},$$

$$\vec{\nabla} \times \vec{E} = -\frac{\partial \vec{B}}{\partial t},$$

$$\vec{\nabla} \cdot \vec{B} = 0,$$

$$\vec{\nabla} \times \vec{B} = \mu \vec{J} + \mu\varepsilon \frac{\partial \vec{E}}{\partial t},$$

and the light was made.

Figure 85. The four Maxwell equations in the classic form [136].

With a mixture of intrigue and fascination, I investigated Maxwell's famous equations, where I discovered that, on the one hand, they did not completely summarise classical electrodynamics and that, strictly speaking, were not by James Maxwell but were rather a compendium of the earlier work of many people like Gauss, Faraday, Ampere and Lorentz [137].

Nowadays, all the physics that allows understanding the universe is known, with a small detail: the whole theory is known, but only from the so-called Big Bang. Early in college, I relentlessly questioned the physics professor: what existed before the big explosion? What rules and conditions were met for such a phenomenon with that amount of unimaginable energy?; the answer was immediate: God; that is to say, there comes a time when one can only say God.

During my academic life, I have had the fortune to meet very special, very valuable people, such as the case of Dr Francisco Bulnes, and I wondered: why is this man with a very high academic level in an institution such as TESCHA[9]?

---

[9] Many educational institutions in Mexico are managed not with the interest of really improving academic quality, or creating deep and basic research, but as simple

Over time, my perception became clearer; meeting him and engaging in small talk in casual encounters within the institution allowed me, on the one hand, to identify an innate investigator in love with science and cognitive development; and, on the other, to awaken in me the need for answers to so many events and phenomena of everyday life. Assertions such as ***"quantum mechanics is the science of God"***, ***"all the knowledge of the universe is also contained in that universe that is your brain"***, and ***"there are three things in life: the truths, the lies and the statistics"*** made me get up unexpectedly from my lethargy and shake myself, as when I was a child. When I told him about my concerns, planted since childhood, he first made it clear that by advancing in science, one realises how ignorant we are, but the good news is that everything can be mathematically modelled.

It surprised me even more that everything interests him, and from then on, from small hallway talks, questions were planted that could become authentic research projects. That said, it seems obvious to emphasise that Dr Bulnes is an excellent mentor, not exclusively in mathematics, but in all aspects that govern the lives of humans. Next, I will quote a brief walk through those corridors, which shows what a great person he is.

On one occasion, observing the work of mobiles developed by electronic engineering students gave rise to investigations that have implications in multiple areas, such as the possibility of autonomous flight ships under purely electromagnetic principles.

In particular, one of our most recurrent talks is related to the human mind and interpersonal relationships because, strangely enough, this mathematician likes contact with people, escaping pre-established roles[10]. The remainder of my participation will be exclusively on these issues.
When I told him about my old concerns about the origins of the universe and what the current mathematical tools are to make these inquiries, he bluntly replied:

> Current mathematical tools are quantum mechanics, quantum electrodynamics and various theories that have been developed by cosmologists, all starting from the ideas of Einstein in new versions trying to unify the microscopic aspects with the macroscopic ones, because curiously everything arises from the microscopic.

---

political platforms for geo-political and economic interests, so it is very difficult to be creative and study science in depth. They are managed by politicians or people very unrelated to academia, or with less diverseacademic interests. Earlier in this book, in Chapter 4, the sad case of education in Mexico was mentioned and analyzed.

[10] Stereotypes.

Continuing with his explanation:

> If we want to know about the entrails of the universe, how our universe is born, what will happen to him, how will his end be; we have to delve into the ideas of the quantum mechanics, they have to be further developed because there is much to be done yet; for example how to reduce uncertainty, since at this time everything is treated through density functions and through a probabilistic framework, something that should not be, in this sense I share Einstein's ideas that everything can be deterministic, but advanced mathematics needs to be developed that included the intention of a field, of course include quantum mechanics.

Another of the talks related to the meaning in the colloquial language of quantum mechanics.

> Assuming that it is the branch of physics that studies nature on small scales, in the fundamental aspect of physics, there are behaviours beyond our understanding at this time, as we have not developed other abilities to understand all the phenomena that arise at the particle level, that is, we need to develop other concepts that have not been developed.

On some occasions, I asked him how he would define conscience, and he immediately and naturally answered:

> Conscience is the active energy of a field, there is always latent energy, but there is an active energy that carries the intention that is the conscience.

His research is focused on field theory, describing the dynamics of physical phenomena at a deep level. Although the initial part of these paragraphs mentions what Galileo and Hawking said regarding the understanding of God's thought, without any doubt, the work of talented mathematicians like Dr Bulnes is to define the rules that shape the language we would use to understand God.

5.6. A meeting of great mathematicians at the Laudal's Fest

Dr Bulnes received an invitation from the mathematics department of the University of Tromso in Norway to attend as a guest lecturer and give a talk on his research in mathematics related to algebraic geometry to celebrate the 80th anniversary of Dr Arnfinn Laudal, a famous world-class professor, and a researcher in algebraic geometry.

Figure 86. Dr Bulnes with Dr Laudal, the great mathematician of the Scandinavian region at the University of Tromso. [138].

Dr Bulnes travelled to Norway for Dr Laudal's fest. Dr Bulnes's talk was entitled "Derived Categories and their Deformed Versions through Integral Geometry Methods II" [139], which was a recapitulation and summary of his latest research results in algebraic geometry on the specific topic of derived categories, their geometric integration over vector sheaves, and field theory in mathematical physics (see Figure 86).

Mathematisches Forschungsinstitut Oberwolfach

Figure 87. Prof. Dr h.c. Gert-Martin Greuel was director of the Oberwolfach Institute from 2002 until 2013 [140]. The Oberwolfach Institute is in the Black Forest and is one of the most important mathematics institutes in the world; all the greatest mathematicians have been through that institute.

At the event, dominated by the attendance of Norwegian and Swedish mathematicians, he met other personalities from the mathematical field, Dr Christian Peskine and the Russian mathematician Dr Victor Palamodov, and significantly Dr Lê Dũng Tráng (Figure 88), who made great contributions to algebraic geometry on singularity theory in the complex domain (*Milnor fibrations, perverse sheaves*).

Figure 88. With Dr Lê Dũng Tráng. Curiously, Dr Lê Dũng Tráng was also a friend of Dr Recillas (Dr Bulnes' doctoral adviser), and here they are celebrating the memory of that great mathematician, Dr Félix Recillas-Juárez [141].

Dr Bulnes made friends with Dr Laudal, with whom he had many conversations during the ten days that the event would last, including talks on a personal level and academia (Figures 86, 88, 89). I believe that Dr Arnfinn Laudal greatly appreciated the attendance of Dr Francisco Bulnes on his 80th birthday.

Figure 89. After the dinner, Dr Bulnes and a group of German mathematicians toast Dr Olav Arnfinn Laudal (centre) in Tromso. Behind Dr Bulnes was Dr Gert-Martin Greuel [138].

That experience for Dr Bulnes was wonderful, and he verified that the great mathematicians have a more holistic vision of the world and share the same concerns of goodwill and generosity for those who want to know more.

….. "God makes them and they come together"……….

As a curious anecdote, during his stay in Norway, Dr Bulnes walked through the centre of Tromso, looking for souvenirs for his family and friends. While he was at the docks, a young Norwegian woman walked with a little boy (around three or four years old). The boy looked at Dr Bulnes, who was looking at the small bay of Tromso (a Norwegian fjord), and he wanted a picture taken with the bay behind him. He asked the young woman to take a photograph, and her son ran and took Dr Bulnes by the hand to also appear in the photo. It seems the boy was looking for a dad (Figure 90).

Figure 90. With a little Norwegian boy in Tromso Bay (Fjord).

CHAPTER 6

THE NANO DEVELOPMENT

6.1. On Nano

We are Dr Helen Thrift and Dr Paul Cladwell, and we will write about Dr Bulnes' research on nanotechnology. Remember that one of Dr Bulnes' postulates and principles in FET is that engineering is creating technology for energy. The universe is all energy because UNISPACE[1] or universal 'covering' is all energy. The global analysis showed that the arguments that use the Lie group's universal covering are within the causality (or causal structure).

According to Dr Bulnes, all this served at the time to define the Hamiltonian dynamics of the universe, useful in characterising the energy states of microscopic entities, such as particles and sub-particles, which are images of the same particle orientated according to their spin. Likewise, there are infinite spins, and the list of sub-particles will be endless. The study of Hamiltonian dynamics requires the study of the energy densities determined by Lagrangians on particles, their momentum and their determined trajectory.

6.2. Conscience Technology on Matter

I am Dr Paul Cladwell from the United Kingdom, a friend of Dr Bulnes. The central idea of Lagrangian dynamics is to study the movement of particles, even in continuous media, and the causes that originate this movement, which brings an intrinsic transformation of space due to energy. Not so the physical entities, which must be invariant under the choice of coordinate system transformations. Dr Fransico Bulnes' idea, after identifying the central idea of Nano as "the power of mind over matter",

---

[1] Unispace is a concept created in mathematical cosmology to explain the unification of the space and time through energy. Likewise, the universe can be seen as all energy entity, where space and time are particles or strings linked by another third entity, energy.

was to characterise these elements of Lagrangian and Hamiltonian dynamic concepts such as the conscience of a particle, due action of a forcefield, the field itself, as intentions to unite the synergistic action of many particles acting together to create an organised transformation on the matter.

These are the first elements considered to create a mathematical nanotechnology theory at the advanced field theory level, which could be achieved in the coming years. Although the latter is mentioned in a first tribute to Dr Bulnes in 2020, this theory is now a reality (Figures 91, 92 and 93).

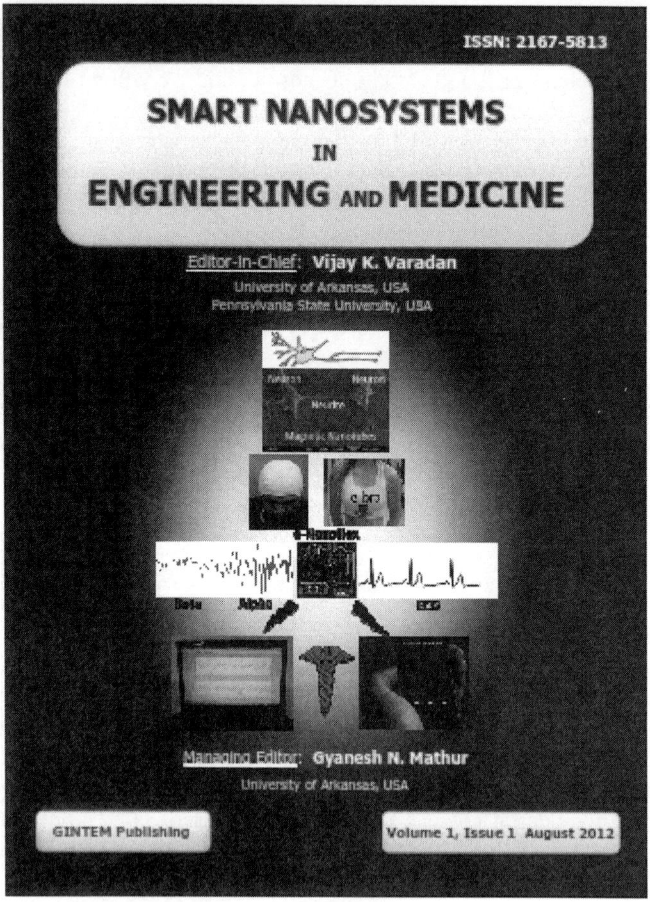

Figure 91. *Journal of Nanosystems*, in the US. Their editor-in-chief is the prestigious neurologist Dr Vijay Varadan of Arkansas University [31, 129].

Dr Bulnes carried out some interesting research in nanomedicine [125-127, 142] (Figure 91) and nanomaterials [61, 143], hinting at what was mentioned before.

This study considers the causal structure of the scattering phenomena through past and future light cones, creating the possibility of energy for one thousand years, perpetual motion machines and star-gates (wormholes in space) with advanced analogues as synchrotronic propulsion (advanced spaceships) and disintegrative mass weapons, using the same principles in all cases.

Dr Bulnes has been developing many ideas around electromagnetic propulsion as a source for take-off and movement for advanced magnetic levitation vehicles, including for landing. In 2010, Dr Bulnes started a research sub-programme of his research programme dedicated to developing advanced vehicles by electromagnetic propulsion (Figure 92).

Several papers have been published in more than 20 journals and book chapters on quantum mechanics and superconductivity [144-145] to prove a consistent theory for the creation of these advanced vehicles and reinforce the study of superconductivity, Majorana fermions, and other related topics (taking advantage of his position as editor-in-chief of the journal on photonics and spintronics, with works by other researchers related to Figure 94).

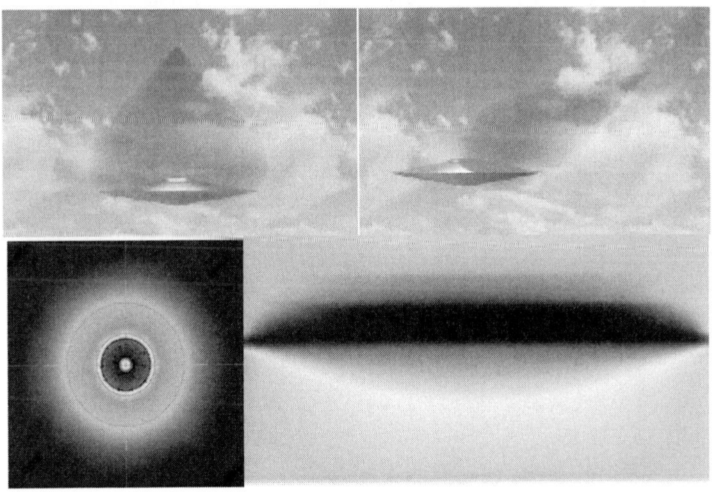

Figure 92. Electromagnetic propulsion research to create an advanced ship with energy for one thousand years. All must be produced from their reactor, considering boson phenomena [31, 144-146].

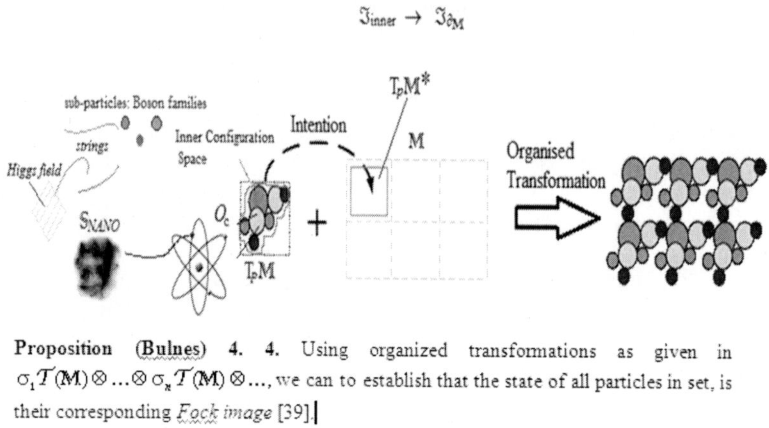

Figure 93. Organised transformations in nanotechnology [132].

Just as charge and gravitational fields cause electromagnetic fields to be caused by weight (mass and force), any rotating objects create torsion fields. Torsion fields can interact with laser beams (change frequency), creating diverse effects, for example, in the biological processes, where torsion directly affects the ADN. Melting or solidifying some materials can increase quartz crystals' properties as resonators and affect some electronic components creating radiation coverings. Torsion can favourably change some liquids and has been noted to affect gravity.

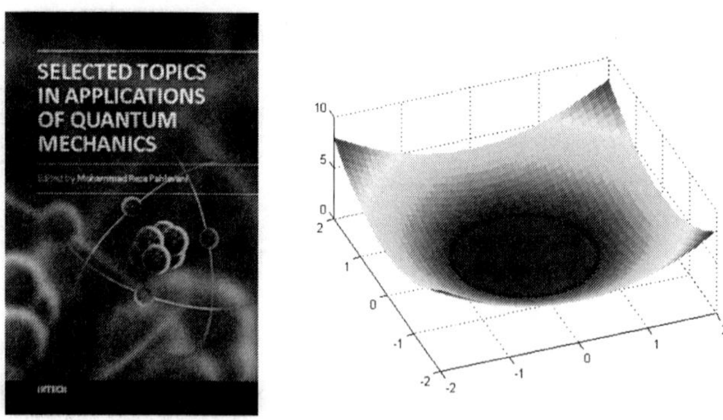

Figure 94. Research on superconductivity and electro-anti-gravitation by Dr Francisco Bulnes [31, 146].

According to this theory, every substance has its own "chronal charge" defined by the quantity of "chronal" particles, which were named "chronons". Dr A.I. Veinik [147] supposed that while an object is spinning, chronons interact with other chronons that surround this object and therefore, the weight of the object changes. According to Dr Veinik's theory, chronons generate the so-called chronal field. Dr Veinik found that strong chronal fields can be generated by spinning masses. Dr Bulnes and M. A. Ramírez [148] measured some properties of chronal fields and found that two types of chronons exist ("plus" and "minus" chronons). It is important to emphasise that the sign of the chronon depended on its spin orientation. Some of these facts could explain the hyperbolicity of space and the laws of minimal action and geometrical trajectories of the "braquistrocrona curve." Satisfying this law of minimal action in the inertial law of Einstein equations with the expansion is reflected in the Christoffel symbols considered in the gravitation equations.

Spinning space can be consigned in a smooth space (as the apparent uniformity of the space-time in the ground) when the energy fluxes of the spins derive in neutrinos, and these fill all the energy spaces. The problem with managing the chronogeometry with these particles is defining the synchronicity required to process to a quantum level and create processes and their integration inside synergic action or "organised transformations" to obtain the reality of space-time [132].

We must create or invent new mathematics to describe complete quantum mechanics based on synchronicity; synchronic quantum mechanics is in the causal structure of the universe but on a deeper level. All these are the ideas of Dr Bulnes around field theory to define the intentionality of a field and apply it in nanotechnology on matter.

6.3. His Mathematical Theory of Nanotechnology: A Short Review

Subsequent research by Dr Bulnes on the intention of a field led him to create a concept within nanotechnology called "conscience", which will be determined by the intention and its action $\mathfrak{I}$ on matter.

In some papers published in many journals of applied mathematics, applied physics, theoretical physics, and specialised journals in mathematics, Dr Bulnes developed and created a theory of nanotechnology that complements some of his modern research theories and his research on derived categories: homotopic theory approaches, Floer homology, Fukaya categories and new integral operator algebras on infinite dimension [149, 150], generalising the concept of Lagrangian in the context of virtual energies that are added to the action $\mathfrak{I}$, to create the concept of the "synergistic action" of a field.

Dr Francisco Bulnes created an axiomatic theory on nanotechnology called mathematical nanotechnology theory [131, 132, 151][2], whose central concepts are the synergic action of a field, the intention of a field, the action of a field, organised transformations, conscience, conscience operators, configuration spaces (defining configuration spaces as $C_{n,m}$), spin manifold, Nano category, and 3-dimensional elemental entity. We can define the intention of a field:

Def. (F. Bulnes) 6. 3. 1. The intention of a field is its active energy.

These brilliant ideas and his research papers on nanotechnology led him to be considered an expert in the field with worldwide recognition.

For Dr Bulnes, nanotechnology should be focused on field theory, where a rigorous review should be made of the concept of the "conscience of a field", which defines the intention to reduce uncertainty in the study framework based on density and probability functions existing in the current development of quantum mechanics. Here is another example where the power of mathematics becomes evident.

6.4. Quantum Algebra for Nano

The space $M_{0,1}$, conforms to a reticular sub-algebra in mathematical logic. Corrective action in memory of the body through a sub-graph Γ, of chords, in the re-composition of the alignment of a vital field in a cybernetic net cube of the body (see figure 95) is shown as:

$$z_j = \int_{x(\Gamma^0)} \theta_1\theta_2\theta_3\theta_4 = \langle [1 \cup \phi(1)] \cup \phi(1) \rangle \cup 1 = 1, \qquad (24)$$

Observe that it can remove the corrective action of encoding memory through a sub-graph

$$:\int_{\Gamma^0} \omega(\Gamma) = 1 \cup -1 \cup 0 \cup [\phi(0) \cup 0 \backslash \phi(1)] \cup 1 = 0 \backslash \phi(1) \cup 1$$
$$= -1 \cup 1 = 1 + (-1) = 0\,, \quad (25)$$

[2] Dr Bulnes published his first essay on this complete theory in [132].

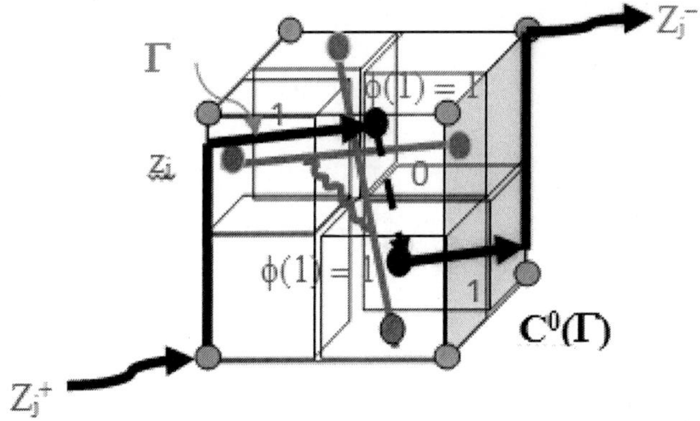

Figure 95. 0000-box of Feynman diagram for the spilling the action of a field [126].

In this case, the cube is the space $C^0(\Gamma)$, into the logic of space-time (Figure 96).

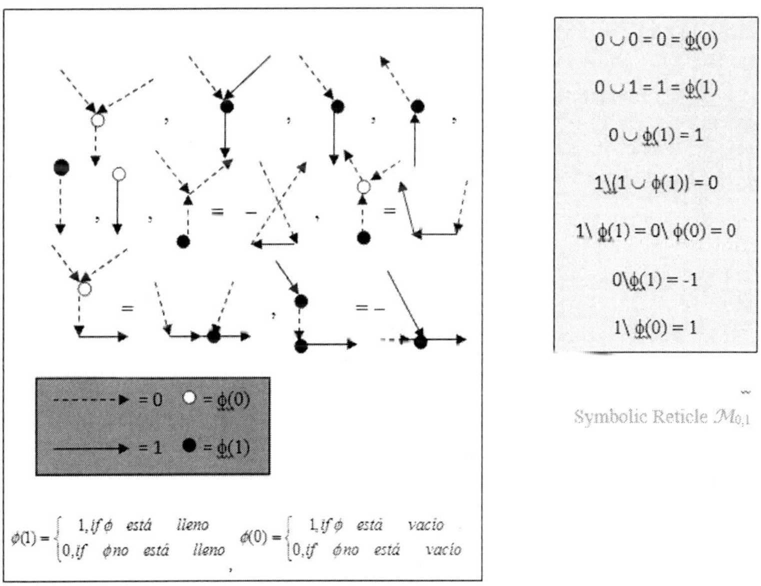

Figure 96. Quantum Algebra for Nano with binary reticle [126, 131].

CHAPTER 7

CULTURAL AND SCIENTIFIC DEVELOPMENT IN THE UNITED KINGDOM AND ITS RELATION WITH DR FRANCISCO BULNES IN RESEARCH AND EDUCATION

7.1 Approach to British Culture and Idiosyncrasies: Its People

This chapter is written by Dr Paul Cladwell and Dr Yuri Stropovsvky. From adolescence, Dr Francisco Bulnes had a special fascination and attraction for British culture, which he learned through books, films, documentaries, and many historical references. Some of its great literary legends wrote about the stereotype of the British as having an ethical sense and being well educated. Dr Francisco Bulnes remembers reading Sir Conan Doyle's novels about the legendary Sherlock Holmes (see Figure 97), the great detective, a possessor of a sharp mind based on logic who solved difficult crimes by observing details, for the police and Scotland Yard. Sherlock Holmes was that stereotype of the British; educated, very observant and with great knowledge acquired by studying many things. He was also combative, fighting hand-to-hand when it was necessary to save a person's life or himself, showing skill in handling various weapons, although Holmes usually did not use them.

Figure 97. A scene where Sherlock Holmes and his assistant Watson both hear the howls of the Baskerville hound on the moors.

He was a regular reader of Charles Dickens, who skillfully portrays Victorian England, masterfully exhibiting reality in the streets of London, including the slums, and the poorest and dirtiest neighbourhoods of the city, despite being a time when the British Empire reached its maximum supremacy colonising the world, and the industrial revolution began. Charles Dickens showed that although the empire lasted a century, it could not eradicate misery and poverty in some sectors of its population.

However, all this amazed Francisco Bulnes more because he understood that wealth sometimes does not permeate the entire population and that, as Dickens puts it, is a problem of the development of consciousness that each individual manages to develop to bring wealth. Many are poor and will continue to be poor because they are poor in spirit. They must steal, prostitute themselves, cheat, and cannot educate themselves.

Francisco Bulnes remembers that during his youth, when he was a student, during his holidays, he liked to read another great author of crime novels, Agatha Christie. He enjoyed the atmosphere of her novels and how descriptive they were, enabling the reader to imagine provincial England, where many of the events took place. Miss Marple, an elderly, conservative English lady fond of solving police cases, had to fight the doubts of the local English police, unlike Sherlock Holmes, who had the support of Scotland Yard.

All of that was fascinating. Dr Bulnes bought many records of English rock music and visited "el Gran Disco", a music shop that sold 33 and 45-RPM vinyl records, which he made a large collection of. His favourites were the Beatles, The Rolling Stones, The Who, Pink Floyd, Yes, Genesis, and many English groups from the sixties and seventies.

British rock and classic British culture from its Victorian era to the 1960s and 1970s, with great changes generated by young people, the arts, the great scientific traditions, and its people with their very phlegmatic behaviour were emerging and made him much more interested in British culture.

7.2. Fine Popular Music, Some Friends, and Approaches to Modern Science Developed in the UK

Francisco Bulnes discovered British rock and its essence while studying for his career as a mathematician. This discovery was so powerful that it made him engage with the British musical movement of the sixties and the first half of the seventies (also incorporating the American one, since the roots and foundations of rock are in blues and the dances and songs of Black culture). However, in a robust, elegant and arguably finer way, rock 'n' roll was strongly expressed in the UK (Figure 98 and Figures 99 A and B).

Figure 98. Yes album cover designed by Roger Dean [152].

A)

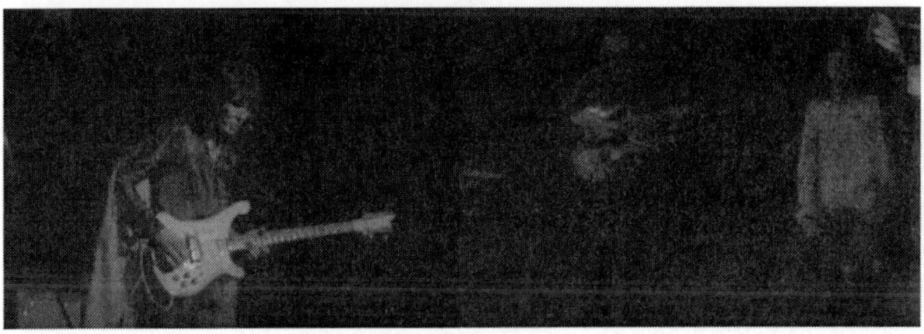

B)

Figure 99. The United Kingdom has had the greatest rock groups in the world. A). The Rolling Stones in 1963. B). Yes live in the UK, Christmas 1972 [152].

Likewise, Francisco Bulnes delved deeper into British culture, developing an interest in learning the language and idioms and understanding British idiosyncrasies. Francisco Bulnes struck up a friendship via regular mail with a young woman named Danielle, exchanging their experiences as young people. It is worth mentioning that this was done before the smartphone and web chat when communication was via letters (see Figure 100). For some time, communication was fluid until it stopped as he was very busy with his studies. In addition, Danielle, who belonged to a very conservative English family, started her bachelor's degree in commerce at the City of London School. After that, she went to live in Manchester.

By then, Francisco Bulnes was a reviewer and adviser on the quality of some scientific journals in the UK. Dr Bulnes received a doctorate honoris causa from the European Union and the United States of America. Lord Hammerton gave him this award (see Figure 101).

This honorary doctorate was awarded to Dr Bulnes for his work in education and as an academic reviewer of several journals indexed in Scopus.

Figure 100. Danielle.

Figure 101. Dr Bulnes at the entrance to the City of London School.

Figure 102. A). Danielle when she met Francisco Bulnes in the UAE. B). Dr Bulnes in the Mahmood Al-Wasimiri's suite being visited by Paul Cladwell and Danielle [153, 154].

Figure 103. Dr Francisco Bulnes at the Houses of Parliament and Big Ben on his first visit to the United Kingdom.

Francisco Bulnes continued his research and bridge-building with the United Kingdom since he was still convinced that it was a very interesting country, full of contrasts that made it particularly attractive (see Figure 103).

He dedicated himself to the study of integral geometry, a mathematical discipline that led him to link Twistor geometry and the integral transform of the latter, attributed to Roger Penrose, and carried out a systematic study of the *D*-algebraic modules, the Radon-Penrose transform, and some deformation theory for the study of string theory. [155].

Dr Bulnes continued to advance scientifically, being a much sought-after reviewer for articles submitted to various scientific journals in Great Britain.

He carried out this activity with great intensity around 2011, when the *British Journal of Mathematics and Computer Science* requested his services as a reviewer [156, 157], including his name within the academic peer review protocol (see Figure 104). Later this scientific journal would change to JAMCS (the *Journal of Advances in Mathematics and Computer Science*) (see Figure 104 and Figure 105).

British Journal of Mathematics & Computer Science
13(6): 1-54, 2016, Article no.BJMCS.19235
ISSN: 2231-0851

SCIENCEDOMAIN *international*
www.sciencedomain.org

Distributional SAdS BH Spacetime-Induced Vacuum Dominance

Jaykov Foukzon[1*], A. Potapov[2] and E. Men'kova[3]

[1] *Center for Mathematical Sciences, Israel Institute of Technology, Haifa, 3200003, Israel.*
[2] *Kotel'nikov Institute of Radioengineering and Electronics, Russian Academy of Sciences, Moscow, 125009, Russia.*
[3] *All-Russian Research Institute for Optical and Physical Measurements, Moscow, 119361, Russia.*

Article Information
DOI: 10.9734/BJMCS/2016/19235
Editor(s):
(1) Yang-Hui He, Professor of Mathematics, City University London, UK and Chang-Jiang Chair Professor in Physics and Qian-Ren Scholar, Nan Kai University, China; Tutor and Quondam-Socius in Mathematics, Merton College, University of Oxford, UK.
(2) David F. Mota, Institute of Theoretical Astrophysics, University of Oslo, Norway.
(3) Morteza Seddighin, Indiana University East Richmond, USA.
Reviewers:
(1) Jos Antonio de Freitas Pacheco, University of Nice-Sophia Antipolis, France.
(2) Francisco Bulnes, Department in Mathematics and Engineering, Tecnolgico de Estudios Superiores de Chalco, Mexico.

Figure 104. An example of the important work of a Russian scientist evaluated by Dr Francisco Bulnes in one of the British journals [157].

More invitations to Great Britain arose, including seminars, classes and talks. In addition, he met wonderful, kind, sensitive British people who were very cultured and knowledgeable about the world. Sometimes Dr Bulnes wondered if these were angels or if they were people sent by them! Dr Bulnes had very pleasant moments in their company.

Britain's tradition in science has been one of the most important, having so many great scientists and scholars like Newton, Maxwell, Dirac, Rutherford, Hardy, Stokes, Green, Cayley, Halley, Faraday, Atiyah, MacDonald, De Morgan, Babbage, Heaviside, Boole, Turing, Penrose, Whitehead, and Hawking, etc. Here we have only mentioned mathematicians and physicists. We can also mention some great mathematicians that, although they were not British, were educated in the universities of the United Kingdom, such as Srinivasa Ramanujan, Harish Chandra and others.

Figure 105. An example of a certificate given to Dr Bulnes for seven years as a reviewer of this and other British mathematics and physics journals. More than two hundred of these certificates were given to Dr Bulnes by this publishing house in the UK [156, 157].

Michelle Musk contacted Dr Francisco Bulnes after her first stay in London to inform him that she had been awarded an honorary doctorate in science, which would be presented to her at a ceremony in Edinburgh the following year [158].

While his attraction and taste for the United Kingdom were still present, Dr Bulnes said that the British were very charming people with a combination of sophistication, education, manners and much culture. Dr Bulnes realised the reputation of the common Briton was tarnished; they are brave and determined. Their education, family values and traditions suggest they read a lot and analyse things and situations. Dr Bulnes feels that the British are ceremonious and accustomed to the standards of humanity and liveliness developed from their history (see Figure 106).

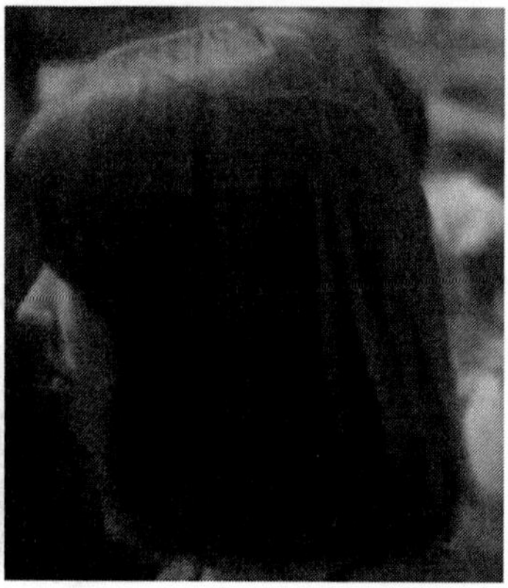

Figure 106. Dr Helen Thrift in Picadilly Circus accompanies Dr Bulnes before his doctorate honoris causa award ceremony. Dr Helen Thrift would be the contact for the prize by the European Union and the USA. While walking, she listens to a rock group in the street with Dr Bulnes.

On another visit to the United Kingdom, he took the opportunity to learn more about the British people, immersing himself in traditions, visiting several English pubs and drinking beer. He took the underground and knew all the cardinal points of the city of London through this transport. He took traditional double-decker buses and went shopping in Piccadilly Circus, where there is a bohemian and cosmopolitan London style (see Figure 107 and Figure 108).

Figure 107. Dr Bulnes in Harrow Mall, where rich British people, artists, actors, and the royal family shop. This place is more exclusive.

Figure 108. Dr Bulnes in Picadilly Circus. Behind him are the famous double-decker buses and black cabs.

A)

B)

Figure 109. A). With the insignia and clothes, Dr Bulnes received the investiture of honoris causa doctorate of Excellence. Dr Bulnes greeting Lord. Peter Hammerton, of Leeds House. B). Dr Helen Thrift when she visited Dr Bulnes in the United Kingdom five years later [31, 159].

Through their expert commission, the European Union and the United States of America decided with the president and vice president of EQAC to reward Dr Francisco Bulnes for his excellent career. This recognition is of great international importance due to the impact of the scientific work of Dr Francisco Bulnes. Likewise, this certifies his great leadership quality (Figure 109).

Subsequently, almost in parallel with her award received by the EU and USA, Michelle Musk (see Figure 110) contacted Dr Bulnes to announce that his name could be included in a compendium of scientists related to the United Kingdom.

Figure 110. Michelle Musk contacted Dr Francisco Bulnes for an interview about his possible candidacy for an honorary Doctorate of Science from the IBC in Cambridge, United Kingdom. Michelle Musk was a follower of Dr Francisco Bulnes because she wanted to be in his biography along with Sara Rains, the editor of biographies of international scientists related to the United Kingdom [158].

Dr Bulnes celebrated his Doctorate of Excellence in an English pub (see Figure 111).

Figure 111. Dr Francisco Bulnes enjoyed dinner in an English pub and had fish and chips with a good beer [31].

Picture 112. Michelle Musk when she was sub-director of IBC in Cambridge, UK.

Life in the United Kingdom can be very hard, and now it has become more so with Brexit (the departure of the United Kingdom from the European Union), and the effects of the pandemic. Michelle had to look for other options besides being the biographer of distinguished scientists (Figure 112). She also is an NYR Organic Independent Consultant.

Michelle Musk has been a good friend to Dr Bulnes. She was born in Cambridge; she knows many legends and histories about Newton's life and other scientists who lived in Cambridge (see Figure 113).

She was always attentive and ready for the requirements of Dr Bulnes regarding information on traditions and folklore in Great Britain.

As a keynote speaker, Dr Francisco Bulnes gave a great talk on symplectic geometry, topology and deformation of derived categories (Figure 114).

Figure 113. Michelle Musk years later, during a Skype interview with Dr Bulnes [157].

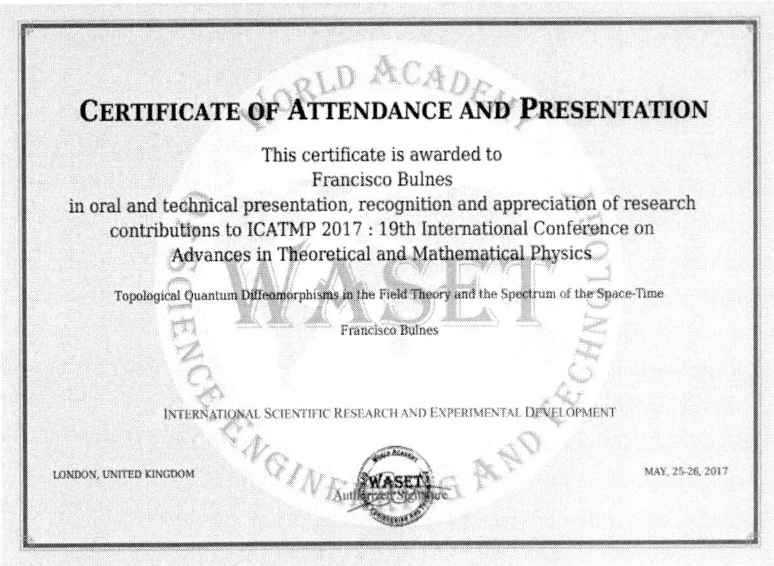

CERTIFICATE OF ATTENDANCE AND PRESENTATION

This certificate is awarded to
Francisco Bulnes
in oral and technical presentation, recognition and appreciation of research contributions to ICATMP 2017 : 19th International Conference on Advances in Theoretical and Mathematical Physics

Topological Quantum Diffeomorphisms in the Field Theory and the Spectrum of the Space-Time
Francisco Bulnes

INTERNATIONAL SCIENTIFIC RESEARCH AND EXPERIMENTAL DEVELOPMENT

LONDON, UNITED KINGDOM — Authorized Signature — MAY, 25-26, 2017

Figure 114. Dr Bulnes gave an impressive talk at an international conference in mathematical physics in London. He was accompanied by Dr Helen Thrift and Michelle Musk [149].

7.3. The Work in the UK Begins to Bear Some Fruit: More travels to the UK

Years of work at the British publishing house as a reviewer, academic adviser or consultant and editor of books (Figure 115), in addition to his educational proposals carried out internationally, led him to obtain several awards from the United Kingdom, the European Union and the United States of America. He has collaborated on several international scientific and research ethics committees in the latter. He has been editor-in-chief for seven years and an editorial board member of many scientific journals on nanotechnology, medicine, engineering and physics too [160, 161].

Figure 115. A compilation of reports and certificates as a British mathematics and physics journal reviewer, including reviews of Commonwealth journals. This book was edited and compiled in the UK [162].

After some years, Francisco Bulnes returned to the United Kingdom to receive several awards, as mentioned before, and other recognition from editorial houses and science publishing houses, such as Cambridge Scholars Publishing, based in Newcastle-upon-Tyne. His stay was for approximately two months.

Figure 116. Dr Bulnes, with Dr Helen Thrift behind him. The photograph was taken by Simon Law, another friend of Dr Bulnes, who accompanied them on the walk. The bridge is over the River Thames.

Simon's knowledge of London made him a great tour guide to show them the sights (see Figure 116).

Figure 117. Dr Bulnes near the House of Parliament on the River Thames.

Simon Law took photos of Dr Bulnes at the Houses of Parliament (see Figure 117). They visited Westminster Abbey (Figure 118), but Simon and Dr Thrift could not go in as they were not British royalty. However, some access was granted to non-British nationals.

Figure 118. Dr Bulnes before entering Westminister Abbey. Westminster Abbey was a monastery in ancient times where the kings and queens of England were crowned.

Dr Bulnes returned to Mexico and then to the United Kingdom in March 2018 to receive his investiture as an Honorary Doctor of Science, according to Michelle's announcement by email and post. He travelled to Cambridge briefly to receive the last recognition of his entire trajectory [163] (Figures 119 A and B).

A)

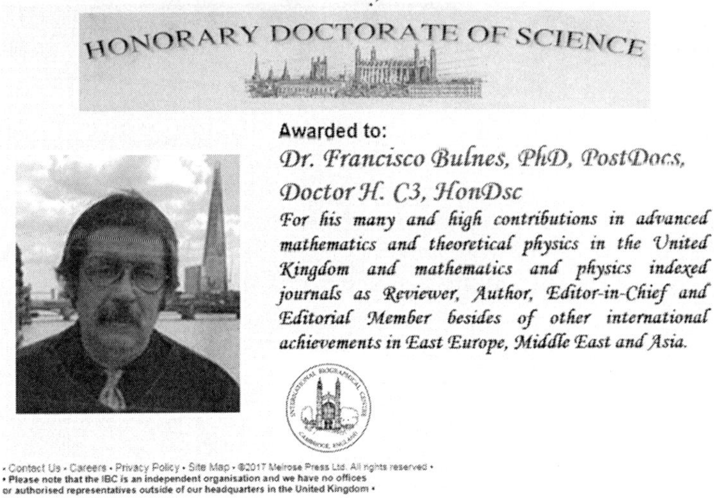

B)

Figure 119. A). Title of Honorary Doctor of Science received in Cambridge, United Kingdom, on the 6th of March, 2018. B). Announced in the media and local newspapers with an explanation about the award.

Dr Bulnes was also invited to assist at a seminar organised by the Isaac Newton Institute for Mathematical Sciences (see Figure 120) in Cambridge, United Kingdom [164].

Figure 120. He visited the Isaac Newton Institute for Mathematical Sciences in Cambridge.

He returned to London from Cambridge and stayed one more week to walk around London and continue to get to know this fascinating city, full of Victorian folklore, modernism and history. Dr Helen Thrift and Simon Law accompanied him. Dr Thrift apologised and only accompanied them on the first day, as she had urgent matters to attend to at Cambridge University. Simon accompanied him on the second day.

Figure 121. Dr Bulnes in front of the River Thames. Across the Thames is the Tower of London, now the Historic Royal Palace, in the distance.

In his desire to learn more about the British people, Dr Bulnes visited bars and restaurants in London. He enjoyed the breeze of the Thames (see Figure 121). On one of his visits, he had his photograph taken with a young "Bobby" (London Police) (see Figure 122) and ate the famous fish and chips on the London docks.

Figure 122. With a "Bobby" (London police). The picture was taken by his friend Simon Law.

Dr Bulnes was always fascinated with this country, the people, and British traditions. He told us: "There is so much history in Britain, which contrasts with its modernism and development in the arts, music, and science".

Figure 123. The Queen Mary, a British battleship used in World War II to support the landing of ground troops. This ship is on the Thames as a museum piece.

He revisited places in London because of the fascination they caused (see Figure 123 and Figure 124).

Upon returning to Mexico, he received recognition and awards from publishing houses and a cash stipend (see Figure 125). He was included on more international scientific and academic committees, and his ranking and data have been included in many academic metrics and scientific classification surveys.

Figure 124. Dr Francisco Bulnes in front of the Tower Bridge with a pigeon.

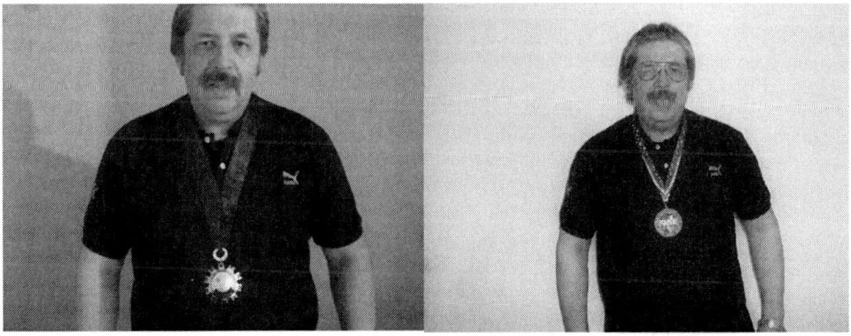

Figure 125. Dr Bulnes with two different medals received due to his brilliant scientific trajectory at different times [164].

The same year in September, he travelled to Germany and received an honorary doctorate, as already mentioned. The following year he received two honorary doctorates, one from the University of Spain-Mexico (see Figure 126) and the other from the Ibero-American Doctoral Senate.

Figure 126. Dr Francisco Bulnes received the doctorate honoris causa from the University of Spain-Mexico. Dr Bulnes gave a speech during the ceremony [165].

Dr Francisco Bulnes published, directed and communicated a collection of mathematical articles by several mathematicians from different countries (see Figures 115, 127).

Figure 127. 2008 luxury edition and volume of mathematics papers collected, directed and advised and Dr Francisco Bulnes in the *Transnational Journal of Pure and Applied Mathematics* [166].

He also received a doctorate honoris causa in mathematics from a subsidiary of Selinus University in the USA, the Auream Phoenix University, and many certificates of appointment as an associate, reviewer and editorial authority of scientific journals and maths journals.

He was invited as an international board member to numerous international events, where his figure was important in attracting scientists and academicians to participate (Figure 128).

Figure 128. Dr Bulnes was invited to the editorial board of many scientific journals in mathematics and physics, nanotechnology and electronic engineering.

Dr Bulnes is still very active, creating more knowledge through high-quality publications, which will possibly be much more valued in the future than now. He continues to focus his research on one of his great passions, the universe, but now with a modern approach to quantum field theory and particle physics (see Figure 129).

Dr Francisco Bulnes founded more research groups with well-defined research programmes and clear goals in field theory and related items.

The academic activity of Dr Bulnes is very intense, from the seminars he directs, his activities as a reviewer and arbitrator of several international events and publications, conferences and presentations in several universities, virtual meetings with members of his editorial board, his work published in scientific journals, to his teaching at the university to which he is attached (Figures 130, 131, and 132).

Figure 129. The 'Hat' Galaxy is one of Dr Bulnes' favourites.

Figure 130. Dr Bulnes was invited to give talks and presentations on his research and latest results at many universities, technological institutes and research institutes.

Figure 131. Dr Francisco Bulnes, accompanied by Dr Sara Gonzalez-Arquímedes, was presented to the Ibero-American Doctoral Senate and awarded a doctorate honoris causa for his brilliant career. This ceremony was in the Old Building of the Senate of the Mexican Republic.

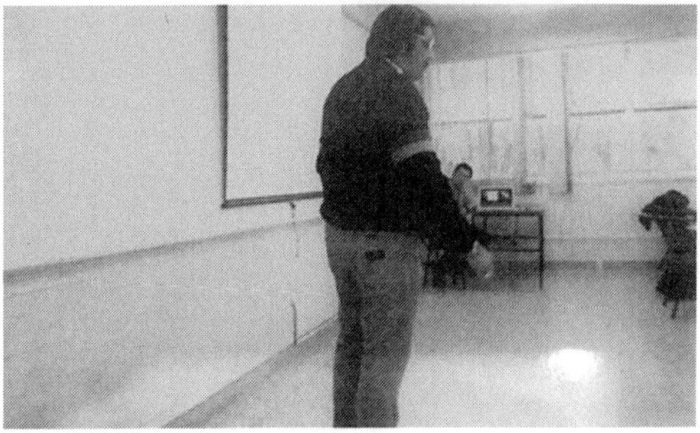

Figure 132. Dr Bulnes giving a talk at the Autonomous University of Queretaro, inside the SMM[1] Congress.

Many countries of the world, such as Germany, Russia, Ukraine, Cuba, the United Kingdom, China, Mexico, Spain, and India, have paid tribute to and published biographies of Dr Francisco Bulnes for his lifetime achievements; his biography has been included in almanacks or compilations of notable and famous scientists and mathematicians. For example, *A Mathematician's Search for Technologies of Understanding the Universe* [31] is the first part of a tribute from the United Kingdom by several scientists and friends that know Dr Bulnes (see Figure 133).

Dr Bulnes has been invited to join numerous international scientific committees and belongs to more than 100 specialised international committees in various areas of science, such as mathematics, physics, nanotechnology, medicine, electronics, electrical and industrial engineering, mechanics and computer science.

---

[1] (SMM), Sociedad Matemática Mexicana.

Figure 133. The first part of a tribute to Dr Francisco Bulnes and his academic, scientific and mathematic trajectory [31].

In this partial biography are included some works and reflections by Dr Bulnes on his research, considering field torsion and curvature energy, nanotechnology, artificial intelligence conjectures and definitions and some notes on his research on superconducting, nanomedicine and his mathematical research theory.

Figure 134. An honorary doctorate is conferred on him by the Ibero-American Doctoral Senate, a ceremony in the Old Building of the Senate of the Mexican Republic on the 7th of June 2019 [167].

Dr Bulnes received more awards and recognition worldwide and locally, and admiration and love from his alums, colleagues, and friends (Figure 131 and Figure 134). He is a charismatic mathematician and scientist.

Figure 135. International accredited speaker from CIC, from the US Award, conferred in 2020 [168].

He has received recognition from many publishing houses for books and scientific journals, including internationally (Figure 135) and commercially.

Figure 136. Award to Dr Bulnes from the *Journal of Electrical Science and Engineering* [169].

Dr Francisco Bulnes has been awarded this recognition for his work as an arbitrator, judge and reviewer for the *Journal of Electrical Science and Engineering (*ISSN: 2661-3247) for 2022, also receiving compensation for his work (Figure 136).

In his deep studies of the universe, he continues to obtain important results and conjectures in the development of his research in field theory (Figure 137).

Electromagnetic waves in conformal actions of the group *SU*(2, 2) on a dimensional flat model of the space-time

Francisco Bulnes

Department in Mathematics and Engineering, TESCHA, Chalco, Mexico

The ultra-hyperbolic wave equation is satisfied using the Penrose integral over $P^1(\mathbb{C})$.

But,considering the Lie group $SU(2,2)$, we can consider conformal theories of gauge fields as electromagnetic fields to measure other fields as gravity [1]. In both directions of a light cone appear the auto-dual Maxwell fields of positive frequency and negative frequency on M (the space-time) respectively that go being added in each time to each orbit. This corresponds to partial waves expansions in 2-dimensions considering the causal structure of the space-time given by these light cones in a 2-dimensional flat model of the space-time [2].

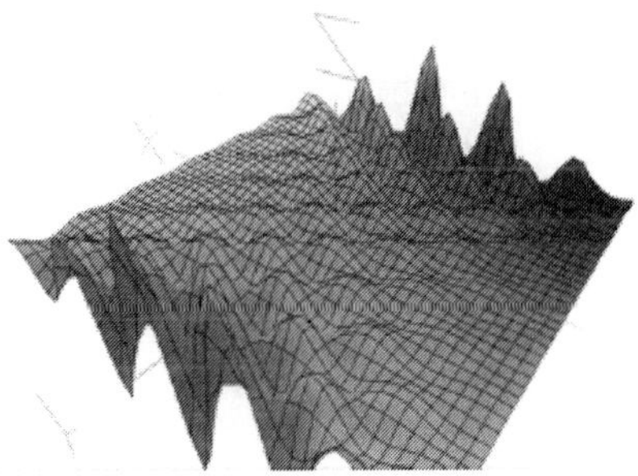

Fig. 1. Electromagnetic waves in conformal actions of the group $SU(2,2)$, on a 2 – dimensional flat model of the space-time. The ultra-hyperbolic wave equation is satisfied. In both sides of axis Y, appear the auto-dual Maxwell fields of positive frequency and negative frequency on M, respectively that go being added in each time to each orbit. This corresponds to partial waves expansions in 2-dimensions.

References

[1] Bulnes F. *Electromagnetic Gauges and Maxwell Lagrangians Applied to the Determination of Curvature in the Space-Time and their Applications*// Journal of Electromagnetic Analysis and Applications, 2012, vol. 4, no. 6, pp. 252–266. DOI: 10.4236/jemaa.2012.46035

[2] Bulnes F. *Mathematical Electrodynamics: Groups, Cohomology Classes, Unitary Representations, Orbits and Integral Transforms in Electro-Physics*// American Journal of

Figure 137. Research on causality structure and conformal actions due to the Lie group $SU(2,2)$, on a 2-dimensional flat model of space-time. This can be generalised to more dimensions of a flat space-time model [170].

Dr Bulnes continues to help young doctoral students achieve their goals (see Figure 138).

Figure 138. Dr Bulnes gives another talk to mathematicians, researchers and young PhD students.

Dr Bulnes has been interviewed on television and has given interviews for many radio channels, both from the Mexican state and private radio stations (see Figure 139 and Figure 140).

Figure 139. "Matemáticas, clave para impulsar el desarrollo sostenible de México"[2], NOTIMEX. Oct 14, 2018, Notimex: http://eitmedia.mx Nota: Si copias y pegas la información de nuestro portal, favor de citar la fuente y agregar el enlace, evita ser sancionado. Derechos Reservados Eitmedia.mx
https://www.eitmedia.mx/index.php/politica/item/25806-matematicas-clave-para-impulsar-el-desarrollo-sostenible-de-mexico [171].

Many of his interviews have been published in major Mexican and foreign newspapers.

Figure 140. Dr Bulnes giving an interview to channel Radio Mexiquense, the most important radio medium in the State of Mexico.

[2] "Mathematics, the key to promote the sustainable development of Mexico".

CHAPTER 8

DR FRANCISCO BULNES'S RESEARCH THEORY

We (Dr Helen Thrift, Dr Michelle Musk and Dr J. P. Cladwell from the United Kingdom) have been studying the mathematical research theories of other research groups, such as IINAMEI, where this research has been applied to the research methodology in all of Dr Bulnes' research, especially in nanotechnology.

In FET, there is an important theorem in the context of the fundamental principle cited by Dr Bulnes: technological actions can be defined through the energy actions of signals of any nature. This is derived directly from the formal definition of engineering. Defining the space of experiments X_{LAB}, as observable sets in a dual Hilbert space where these experiments are realised on a true proposition set, is the same as the creation of technologies on the principle of a true proposition set.

Theorem (F. Bulnes) 8. 1. [14] A). X_{LAB}, is isometric to E_{FISMAT}, B). $\Phi_\alpha(t_\beta) = X_{LAB}(\Phi_\alpha)$.

Proof. Its demonstration can be found in [132]. *A grosso moda*, the theorem says A) that X_{LAB}, and E_{FISMAT}, follow the same metric of acknowledgement. An example is the research line given for curvature sensors.

Incise B) says that the results in a laboratory must be congruent with the results in theory. Furthermore, establishing an identity inside the knowledge creation (epistemology) is an equivalence because much knowledge can be produced with improvements to existing technologies. Experimental research must be founded on the theory of knowledge. Both are measurable in duality.

This generates certainty and consistency in the research perspective [66], as affirmed by Dr Bulnes. The field observables are elements of H^*, where H^*, is the 2-curable measure of energy signal space[1].

---

[1] We consider the mapping $T: H^* \to H^*$, with rule of correspondence $f(x) \to T(f(x)) = \mu(x)$, [18, 19].

In modern research, managing time is fundamental. Likewise, established priorities classify the research projects and lines concerning time as large, average and short-scope projects.

The research time can be designed as time series, outlining the objectives in steps. This obeys the proper research evolution. What defines that evolution?—research development. How can we establish that this research development is the most optimal? It obeys the efficient causality principle in the universe, which affirms that any optimal or efficient effect must come from an efficient cause. Then an optimal design of the cause is necessary. Here is where it starts.

Time is the shortest length between cause and effect.

Proposition (F. Bulnes) 8. 1. The efficient cause inside research theory comes from a sound theory.

Proof. See [17].

All applications will be corollaries of this theory. Then, theoretical knowledge is fundamental. However, we also have a universal principle that the cause is included in each one of the applications because the effect is included in the cause, and the cause is included in the effects. The great mathematician Hilbert (the modern Euclid) said: "The best application of the acquired knowledge is a good theory."

In the research method for fundamentals, causality is given in the design of a $Th\Sigma$ consolidated through publications in indexed and arbitrated journals and reviewed by academic peers. Likewise, the research ethic is normal for the rules established to accredit the scientific work. Then we can define:

Def. (F. Bulnes) 8. 1. Ethics in research are the moral quality or harmony between the individual and scientific work.

Research problems related to the design, creation and functioning of a transducer were solved. Then in the context of the technological prototypes, we considered the following commutative diagram [32, 44]:

$$
\begin{array}{ccccc}
Th_{App} & \rightarrow & SubTh\,\Sigma & \rightarrow & \Psi_{\alpha} \\
\downarrow & & \downarrow & & \downarrow \\
\{\pi_j\} & \rightarrow & \pi^* & \rightarrow & \Psi, \forall\, j \in [\alpha] \\
\downarrow & & \downarrow & & \downarrow, \\
\{\pi_K\} & \rightarrow & \pi^* & \rightarrow & P \\
\downarrow & & \downarrow & & \downarrow \\
\{\pi_K\} & \rightarrow & \pi^* & \rightarrow & P
\end{array}
$$

where Ψ_{α}, is the model's algebra of the class α, the set $\{\pi_j\}$, is a set of prototypes π_K, to develop through the application of a theory $Th_{App} \rightarrow SubTh\Sigma$, the optimal prototype π^*.

This optimal final prototype is a product that can solve a social problem: $P \rightarrow 0$.

In the same year, 2017, the executive assistant of IBC, Michelle Musk (see Figure 112), communicated to Dr Francisco Bulnes the intention to include his biography in the first edition of a book (published in the summer of 2018) *Great Men and Woman of Science* [141]. After a hard, scrupulous and detailed study, she told him several publishing houses would publish his great academic citations.

Figure 141. Book on reviewers and authors published and revised in British mathematics, physics and technology journals. This book reported more than 100 revised papers. All contents are registered with Publons [156, 162]. B) and published by Melrose Place Press, Cambridge, UK [171]. The editor of this book is Sara Rains.

Dr Bulnes went on a short trip to Cambridge University after talking to some mathematicians such as Townsend and Wilson. In a small ceremony inside

Trinity College, they received a biographical compendium of the most distinguished scientists in the world, but related to the United Kingdom (see Figures 141, 142 and 143).

Figure 142. Trinity College Great Gate. Trinity College is part of the University of Cambridge, United Kingdom.

Figure 143. Dr Bulnes at Trinity College Great Gate, in Cambridge, UK.

Continuing with the mathematical theory of research, Dr Francisco Bulnes proposed a block diagram for fundamental research based on his ideas from the results we have previously analysed in this chapter.
Imagine we want to manufacture a special rod with maximum resistance tension with minimal displacements $u(t)$, inside. The special rod could be technology t_σ, which we call t_α, belonging to class α of beams and pillars. First, as theoretical research, Bulnes's research method can be described as research types 15, 44, and 172 (Figure 144):

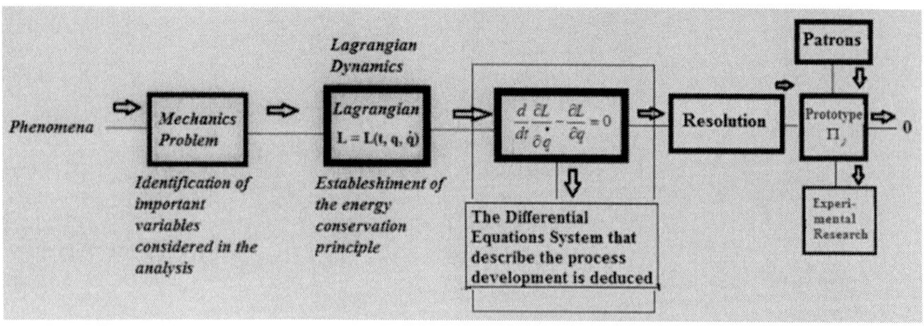

Figure 144. Blocks diagram of research in physical dynamics.

After the prototype level $Th_{App} \rightarrow SubTh\Sigma$, we have its evaluation given for the value of the execution integral:

$$Exe(t_\sigma) = \int_{\Phi_\alpha} \varphi_\gamma \, \Phi_\alpha A(\Phi_\alpha(E_{FISMAT}))\mu(\Phi_\alpha E), \quad (26)$$

which will derive from an integral of energy,

$$\|u\|_{Pattern}^2 = \int_0^1 GS(t)u'^2\,dt$$

$$\leq ES_1 \int_0^1 \left(u^2(t) + u'^2(t)\right) dt \leq GS_1 \|u\|_{1,2}^2, \quad (27)$$

where G is Young's modulus, $S(t)$ is the area of the cross-section with $S_0 \leq S(t) \leq S_1$, and $u(t)$ is the displacement of the cross-section of the rod at point
t, in the longitudinal direction. We have supposed that $t = 0$, and is fixed, that is to say, $(0) = 0$.

It is worth mentioning that using his rigorous method, with the principles of functional analysis developed in his mathematical theory of research, Dr Bulnes has developed a pair of transforms in materials science for the quasi-relaxation problem. This transforms pair is:

$$\Psi(\tau) = QX\{\phi(t)\} = \int_0^\infty \phi(t)e^{-\frac{t}{\tau}}\,dt,$$

$$\phi(t) = QX^{-1}\{\Psi(\tau)\} = \int_0^\infty \Psi(\tau)e^{-\frac{t}{\tau}}\,d\tau,$$

Figure 145. Promotional poster for Dr Bulnes' mathematical research theory.

CHAPTER 9

IN HIS COMMUNITY

Despite being a rigorous academic, a renowned mathematician and a scientist, Dr Francisco Bulnes is very charismatic, achieving admiration from all the people in his community. His students love and admire him, always referring to him as an example to follow. Teachers respect and affectionately admire him. The community generally respects him and considers him a man of great honour. Some people in his community have made friends with him, joining those great friends he has in his life, Mr Pedro Alejandro, his wife María de los Ángeles, his son Manuel Alejandro, and his daughter Jaqueline Alejandro; he appreciates them and holds them in high regard. Mr Pedro Alejandro works tirelessly for the community's well-being.

Mr Damián Lugo and his son Damián are natives of Amecameca and have been forest guards of the Popocatepetl Volcano for many years. They are good friends of Dr Francisco Bulnes.

His best friend Jorge Calles-Fuentes owned El Gran Disco, the most famous record shop (1965–2003), selling 50s, 60s, 70s, 80s and 90s records.

Many alums (such as Omar Zamudio, Andy Rodrigues, Itzcoatl Valencia, Cuauhtémoc Gutierrez, Victor Sánchez-Suárez, Luis Alfredo Ortiz-Dumas, and Brandon Martínez) and academic fellows have been his friends. Dear friends such as Isaías Martínez Palma, a great electronics engineer (someone full of light) and his family (his daughter Jatsiri, his wife María del Pilar Hernández, and his son Isaí Mahonri), Rocío Cayetano (the warrior teacher), Master of Science J. C. García-Limón (the best electronics engineer), and Jorge Tello. Other great friends are Roberto Galicia, a brilliant chemical engineer, and José Galicia, "Dr Bulnes's endearing brothers" (his brothers-in-law).

He has friends in public administration like Rodolfo Martínez-Calles (formerly finance director of TESCHA). He remembers his friends Luis Motta-Adalid and Jaime García-Bravo, who were like brothers in past epochs.

CHAPTER 10

SAVE OUR EARTH

In recent years, the deterioration of the Earth and its natural environments, caused by the contamination of the air, rivers and our soil, has led Dr Francisco Bulnes to conduct more research focused on proposals for technologies based on field theory.

The field theory approach not only proposes theories to explain the entire phenomenology of the universe or specific issues of some technologies but also that these technologies must be developed with their natural patterns, designed with the best optimisation for clean energy from microscopic sources, with the development of advanced transducers designed with nanotechnology and quantum electronics.

Michelle and Helen said that Dr Francisco Bulnes established the fundamental principle to carry out any technological innovation in the world, using only quantum field theory and everything that goes with it.

Under the principles of this great theory, we will no longer damage our home, which is the Earth, and we will also be able to advance processes and transformations that until now have seemed like science fiction, but as Dr Bulnes says, they are achievable thanks to QED (quantum electrodynamics) and nanotechnology. This has to be used under the mathematical principles of field theory developed with mathematical physics, symplectic geometry, differential topology, dynamical systems and global analysis.

Today more than ever, we must change our way of life, our individual and collective perception of the true value of things, our perspective on the most valuable, and our understanding that we are all passing through and will take nothing with us. Leaving much money behind leaves problems and foments greed. Leaving much knowledge behind leaves solutions to problems and fosters the growth of our consciousness.

Currently, we continue to think like primitives. The only thing that has changed is technology and comfort for a few. But essentially, we are the same; sadly, there has been no evolution because there has been no growth of consciousness that has managed to move something inside us and cause a *new awakening*.

The world must be guided by those with greater knowledge and acquired wisdom, not by the rich who are sick with greed or politicians who are sick with power. Both are sick and corrupt entities, so humanity must ignore them and be guided by wisdom.

TECHNICAL NOTATIONS

$\kappa(\omega_1, \omega_2)$–spectral curvature or curvature energy. This is given in $\frac{V}{m^3}$, [23].

$\mathbb{R}^3$ – real ordinary space or 3-dimensional real space.

$\tau(\omega_1, \omega_2)$–torsion energy or spectra of torsion. Also given in $\frac{V}{m^3}$, [26].

Ψ_{α_i},–model algebra of the ith –class α_i. For example, all models of the aeroplanes of glider type of the class airbus.

M –space-time model as 10-dimensional space constituted as a manifold of a complex Riemannian manifold structure.

$\{\pi_j\}$–prototype sequence or succession.

H^*–Dual Hilbert space. This is the Hilbert space of measures of integrable squares, for example, the registers on an instrument panel.

$\mathrm{D}M_{(.)}^{-}(k, R)$–tensor triangulated derived category of a certain type (.) on field

k, a ring of modules R.

$\Omega(\Gamma)$–space of possible trajectories or paths in space-time that can follow a particle. $\Omega(\Gamma) \subset \mathbf{M}$, on all the elections of possible paths whose statistical weight corresponds to the determined one by the intention of a field.

$L^2(O_c(z(t), \mu))$ –Measures the space of all electronic propagators of all possible paths that can take the cognitive process of accumulated experiences in the memory of an intelligence. This space has a Hilbert space structure.

$L^2[\phi]$ –space in functional analysis. This has a Hilbert space structure. In dynamics and quantum mechanics, this space is the Hamiltonian space.

$\mathfrak{I}$ –field action.

$\phi_i(i = 1, 2, 3, \dots\dots, n)$ – ith –state of energy of a particle or body.

ω_i – ith –1-differential form in complex geometry.

M –Riemannian orientable compact manifold.

O_c –conscience operator.

Ω_L –Chern-Simmons form.

$\mathfrak{R}$ –equivalence relation.

$\omega(\phi(x))$ –volume form of the energy state $\phi(x)$.

I –exploration operator.

G −topological group of the operators class of acknowledge. A topological group in the invariance of correspondences in the derived categories.
$D_F(x(t))$ −electronic propagator of the particle x, with trajectory $x(t)$.
$C_{n,m}$ −configuration field space.
$\int d(\phi(x))$ −quantised trajectories.

_II$(\rho, \boldsymbol{v})$ −topological vector space of a fluid of density ρ, and velocity v.
.$M_h(G, \mathbf{C})$ −Hitchin moduli space equivalent to space-time in geometrical correspondences and ramification theory.
P_N −complex projective space of dimension N.
Ω_{ij} −differential forms of curvature in a Kähler manifold.
D − modules − spaces of differential operators that are elements (germs) of an algebraic sheaf. These modules belong to a twisted sheaf of differential operators on a smooth algebraic variety.
K_μ^{ab} −contortion tensor. It is a curvature in the torsion sense.
$CP - Violation$ −in particle physics, *CP* violation is a violation of *CP*-symmetry (or charge conjugation parity symmetry): the combination of *C*-symmetry (charge conjugation symmetry) and P-symmetry (parity symmetry).
$GUT - models$ −grand unified theories. These theories are being developed to explain all aspects of strong interactions, weak interactions, supergravity, etcetera.
$P - symmetry$ −symmetry due to the parity.
$C - symmetry$ −means the symmetry of physical laws under a charge-conjugation transformation. Electromagnetism, gravity, and strong interactions all obey *C*-symmetry, but weak interactions violate *C*-symmetry. The charge conjugation is a transformation that switches all particles with their corresponding antiparticles and thus changes the sign of all charges: electric charges and those relevant to other forces.
νMSM −a minimal re-normalisable extension of the standard model by right-handed neutrinos. This model explains the neutrino oscillations and provides a candidate for dark matter and a mechanism of baryon number generation in the early universe.
$SO(3)$ −rotation group of range 3. Their elements are the matrices in 3-dimensional real Euclidean space: $A = \begin{pmatrix} cos\theta & -sin\theta & 0 \\ sin\theta & cos\theta & 0 \\ 0 & 0 & 1 \end{pmatrix}$, $\forall\theta \in [0, 2\pi]$. This group is a Lie group.
$SO(4) - a$ group of rotations about a fixed point in four-dimensional real Euclidean space. Also is a Lie Group.

$SU(2) - a$ special unitary group of degree 2, which is the Lie group of 2×2 unitary matrices with determinant 1. $SU(2) \subset U(2) \subset GL(2, \mathbf{C})$.
_II$(\sigma, \boldsymbol{E})$ −topological vector space of a fluid of conductivity σ, and electric field $\boldsymbol{E}$.
$\mathbf{C}$ −complex numbers space. This is the set of points z, such that $z = a + ib, \forall\, a, b \in \mathbf{R}, and\ i = \sqrt{-1}$.
C −algebraic curve. This is an element of a moduli space in algebraic geometry. Field theory has the re-interpretation as an element of string theory corresponding to an operator in QFT, also used as a complex.
$\overline{\psi}\gamma_5\gamma^{\mu}\psi$ −term of spinor axial current.
$\kappa J^{\mu}_{(A)} J_{(A)\mu} - a$ term that refers to the fermion self-interactions induced by quantum torsion.
$I_{NY} - a$ term, which is the Nieh-Yan topological invariant density which creates the unique Lorentz invariant torsion.

ACRONYMS AND NOMENCLATURE

AMS—American Mathematical Society. It is the quality and subject classification standard in mathematics aims and scopes.
The UNAM—National Autonomous University of Mexico. This is the principal and most important university in Mexico.
GDP—gross domestic product. The standard measure of the value added created through producing goods and services in a country during a certain period. As such, it also measures the income earned from that production or the total amount spent on final goods and services (less imports).
IM-UNAM—Institute of Mathematics of the UNAM.
ESIME-IPN—Superior School of Mechanical and Electrical Engineering of National Polytechnique Institute. This university is important in Mexico.
ESFM-IPN—Superior School of Mathematical Physics of National Polytechnique Institute.
CINVESTAV—Centre of Advanced Research. A centre in Mexico dedicated only to postgraduates and research.
IAS-Princeton—Institute of Advanced Studies of Princeton.
CEPAL (Comisión Económica para América Latina y el Caribe)—Economic Commission to Latin America and Caribbean Region.
UPN—National Pedagogic University.
TFI—Teoría Formal de la Ingeniería.
OECD—Organization of Economics and Commerce Development.
PEMEX—Mexican Oil.
WB—World Bank.
CONACYT—Science and Technology Council.
UJAT—Universidad Juárez Autónoma de Tabasco. Jarez Autonomous University of Tabasco.
TFT—topological field theory. This is a quantum field theory which computes topological invariants.
QFT—quantum field theory. This studies aspects of the quantum field.
QED—quantum electrodynamics. This is a theory of particle physics, the relativistic quantum field theory of electrodynamics. It describes how light and matter interact and is the first theory where full agreement between quantum mechanics and special relativity is achieved. QED mathematically describes all phenomena involving electrically charged particles interacting using an exchange of photons and represents the quantum counterpart of

classical electromagnetism, giving a complete account of matter and light interaction.
BRST-Cohomology—in theoretical physics, is a BRST formalism, or BRST quantisation (where the BRST refers to the last names of Carlo Becchi, Alain Rouet, Raymond Stora and Igor Tyutin) that denotes a relatively rigorous mathematical approach to quantising a field theory with gauge symmetry.
UNISPACE—this concept was created in mathematical cosmology to explain the unification of space and time through energy.
FET-formal engineering theory. English acronym of TFI.
SMM—Mexican Mathematical Society.

REFERENCES

[1] Mathcmatics Genealogy Project, n.d., Félix Recillas-Juárez, PhD thesis, https://www.genealogy.math.ndsu.nodak.edu/id.php?id=40036

[2] Mathematics Genealogy Project, n.d., Francisco Bulnes, PhD thesis, https://www.genealogy.math.ndsu.nodak.edu/id.php?id=142573

[3] Francisco Bulnes, 2000, *Teoría de los* (g, K) *−módulos*, Monograph of Instituto de Matemáticas, UNAM, Ciudad Universitaria, CDMX.

[4] Francisco Bulnes, 2001, *Conferences de Matemáticas I: Seminario de Representaciones de Grupos de Lie Reductivos Reales*, Félix Recillas Juárez (Ed.), Instituto de Matemáticas, UNAM, Ciudad Universitaria, CDMX.

[5] Francisco Bulnes, 2002, *Conferences de Matemáticas II: Seminario de Representaciones de Grupos de Lie Reductivos Reales*, Félix Recillas Juárez (Ed.), Instituto de Matemáticas, UNAM, Ciudad Universitaria, CDMX.

[6] Francisco Bulnes, 2005, *Conferencias de Grupos de Lie*, Colección de Matemáticas Avanzadas Appliedmath 1, IMUNAM, IPN-ESIME, IPN-ESFM, SEP, CDMX.

[7] Francisco Bulnes and M. Shapiro, "On General Theory of Integral Operators to Analysis and Geometry," IM/UNAM, SEPI/IPN Monograph in Mathematics, 1st Edition, P. Cladwell, Mexico, 2007.

[8] International Research Group in Mathematics and Electronics, n.d., https://medium.com/@liuxortiz/international-research-group-in-mathematics-and-electronics-49575928dc71.

[9] Grupo de Investigación TESCHA, (2009-2015), Reg. Comecyt, Club de Ciencias 5296.

[10] F. Meneses-Tello, El legado de Félix Recillas Juárez a los estudiantes de Matemáticas: Colección de Libros de Matemáticas para el IM/UNAM, Matemáticos en México, http://matematicos.matem.unam.mx/matematicos-r-z/matematicos-r/felix-recillas/803-el-legado-de-felix-recillas-juarez-a-los-estudiantes-de-matematicas

[11] Francisco Bulnes, 2014, Integral geometry methods on deformed categories to geometrical Langland's ramifications in field theory, *Ilirias Journal of Mathematics*, 3(1), 1-13.

[12] Francisco Bulnes, "Moduli Identities and Cycles Cohomologies by Integral Transforms in Derived Geometry," *Theoretical Mathematics and Applications*, vol.6, no.4, 2016, 1-12.

[13] Francisco Bulnes, Integral Geometry Methods in the Geometrical Langlands Program, SCIRP, USA, 2016. ISBN: 978-1-61896-139-6.

[14] Francisco Bulnes, "Prospective and Quality of Scientific and Technological Research," SEP, TESCHA, State of Mexico Government, 2010.

[15] Francisco Bulnes, Teoría de la Investigación en Ciencias de la Ingeniería, EAE, Alemania-España, 2013. ISBN: 978-365-906-879-9.

[16] Francisco Bulnes, "Integral geometry and complex integral operators cohomology in field theory on space-time," in Proceedings of 1st International Congress of Applied Mathematics-UPVT (Mexico), 2009, vol. 1, Government of State of Mexico, pp. 42-51.

[17] Francisco Bulnes, R. Goborov, "Integral Geometry and Complex Space-Time Cohomology in Field Theory," *Pure and Applied Mathematics Journal*. Special Issue: Integral Geometry Methods on Derived Categories in the Geometrical Langlands Program, Volume 3, Issue 6-2, pp30-37.

[18] Francisco Bulnes, "Mathematical Electrodynamics: Groups, Cohomology Classes, Unitary Representations, Orbits and Integral Transforms in Electro-Physics," *American Journal of Electromagnetics and Applications*. Volume 3, Issue 6, November 2015, pp. 43-52. doi: 10.11648/j.ajea.20150306.12

[19] Francisco Bulnes, *Curvature of a Universe and Radon Transform*, PostGraduate Thesis, Faculty of Sciences, UNAM, 2001.

[20] Francisco Bulnes. Research on Curvature of Homogeneous Spaces, 1st ed.; TESCHA: State of Mexico, Mexico, 2010; pp. 44-66.

[21] Francisco Bulnes. Martínez, I, Mendoza, A. Landa, M., "Design and Development of an Electronic Sensor to Detect and Measure Curvature of Spaces Using Curvature Energy," *Journal of Sensor Technology*, 2012, 2, pp116-126. http://dx.doi.org/10.4236/jst.2012.23017.

[22] Francisco Bulnes, Isaías Martínez, Omar Zamudio, Gabriel Negrete. "Electronic Sensor Prototype to Detect and Measure Curvature Through Their Curvature Energy." *Science Journal of Circuits,*

Systems and Signal Processing.Vol.4, No. 5, 2015, pp. 41-54. Doi 10.11648/j.cssp.20150405.12.

[23] Francisco Bulnes, I. Martínez, O. Zamudio, "Fine Curvature Measurements through Curvature Energy and their Gauging and Sensoring in the Space, Advances in Sensors" Reviews 4, (Ed.) Sergey Y. Yurish, IFSA, 2016.

[24] Francisco Bulnes, "Dual representation of the curvature in a Hilbert space: curvature and integral transforms," *JP Journal of Geometry and Topology* 26(1) (2021), 39-51. DOI: 10.17654/GT026010039.

[25] Francisco Bulnes, "Electromagnetic Gauges and Maxwell Lagrangians Applied to the Determination of Curvature in the Space-Time and their Applications," *Journal of Electromagnetic Analysis and Application*, Vol. 4 No. 6, 2012, 252-266. DOI: 10.4236/jemaa.2012.46035.

[26] Francisco Bulnes, A J. C. García-Limón, L. A. Ortiz-Dumas, V. A. Sánchez-Suarez, "Detector of Torsion as Field Observable and Applications," *American Journal of Electrical and Electronic Engineering*. 2020, 8(4), 108-115. DOI: 10.12691/ajeee-8-4-2.

[27] Francisco Bulnes, Stropovsvky, Y. and Rabinovich, I., 2017, "Curvature Energy and Their Spectrum in the Spinor-Twistor Framework: Torsion as Indicium of Gravitational Waves." *Journal of Modern Physics*, 8, 1723-1736. doi: 10.4236/jmp.2017.810101.

[28] Francisco Bulnes, 2017, "Detection and Measurement of Quantum Gravity by a Curvature Energy Sensor: H-States of Curvature Energy, Recent Studies in Perturbation Theory," Dr Dimo Uzunov (Ed.), InTech, DOI: 10.5772/68026. Available from: https://www.intechopen.com/books/recent-studies-in-perturbation-theory/detection-and-measurement-of-quantum-gravity-by-a-curvature-energy-sensor-h-states-of-curvature-ener

[29] Doctor Honoris Causa for the Mathematics and Conscience, 2021, https://www.facebook.com/holaODAEE/photos/a.115854898441998/4670461836314592/?type=3

[30] Cote, D., 2017, *Matemático Riguroso y sus Teorías para Tecnologías: Análisis y Reflexión sobre un Matemático Iconoclasta*. EAE, Spain-Germany. ISBN-10: 6202236531, ISBN-13: 978-6202236539.

[31] Helen Thrift, Yuri Stropovsvky, Francisco Bulnes (Honored), *A Mathematician's Search for Technologies of Understanding the Universe*, Cambridge Scholars Publishing, Newcastle, United Kingdom, 2020. SBN: 978-1-5275-5544-0.

[32] Bulnes, *Zuckerman Functors*, Compilación de Ponencias del Seminario de Teoría de Representaciones de Grupos de Lie Reductivos Reales, Instituto de Matemáticas, UNAM, 2003.

[33] Doctoral Seminar of Representations Theory of Real Reductive Lie Groups, Schedule of Lectures and Talks. Instituto de Matemáticas, UNAM, Ciudad Universitaria, 2002.

[34] Photographs of particular Archives, Dr Francisco Bulnes on his trips on South Eastern of Mexico.

[35] Photographs and Report Archives of UTECAM, AMIME, PEMEX, UVM-Villahermosa, on visits realised by Dr Francisco Bulnes to Coatzacoalcos, Veracruz, Villahermosa, Tabasco y Ciudad del Carmen, Campeche, 2002-2006.

[36] Francisco Bulnes, "Mukai-Fourier Transforms as Solutions to Field Equations: Spectrum as Higgs-Oscillations of H-States in the Space-Time," *Topological Methods in Dynamics and Related Topics. Shilnikov Workshop*, Nizhny Novgorod 12-13 December 2020, Russia.

[37] Francisco Bulnes, "Characterised cycles integration on D-modules as solutions in L-holomorphic bundles," AGMA (Algebraic and Geometry Methods of Analysis), May 24-27, Odesa, Ukraine, 2022.

[38] Francisco Bulnes, Contemporary Mathematics. Fundamental Directions (Современная математика. Фундаментальные направления), RUDN University via the Russian Academic Excellence Project, DAAD via the *German-Russian* Interdisciplinary Science Center, Russian Foundation for Basic Research,. 2017.

[39] Francisco Bulnes, n.d., "Mukai-Fourier Transforms as Solutions to Field Equations in Derived Categories: Spectrum as Higgs-Oscillations in the Space-Time V," *Programme f International Seminar* OTDE "Operator Theory, Differential Equations and their Applications" North Caucasus Center for Mathematical Research of the Vladikavkaz Scientific Center of the RAS, Southern Mathematical Institute of the Vladikavkaz Scientific Center of the RAS, Russia.

[40] Research Proceedings, Appliedmath III, Dr Francisco Bulnes-Aguirre (Ed.), IPN, UNAM, CINVESTAV, UAM Tec-Monterrey, 2007.

[41] Seyed Mohammad Ajdani, Francisco Bulnes, "Vertex Descomposability of K-Cohen-Macauley Simplicial Complexes of Codimension 3," *Advanced Mathematical Models & Applications*, Vol.5, No.3, 2020, pp.303-306.

[42] Seyed Mohammad Ajdani and Francisco Bulnes, Algebraic properties of the path complexes of cycles, *International Journal of Advances in Mathematics*, Volume 2019, Number 5, 1-8, 2019.

[43] Francisco Bulnes, *Notas Curso de Prospectiva y Calidad de la Investigación Científica y Tecnológica Internacional*, Curso de Investigación, 2011.

[44] Francisco Bulnes, El Modelo de Competencias en la Investigación Moderna, EAE, Alemania-España, 2018. ISBN: 978-620-2-13528-3.

[45] Francisco Bulnes, "La Verdadera Formación en la Educación: Personas hacia el Pensamiento Crítico, Congreso Online Aportes de Paz al mundo en transición y post COVID-19,'' ODAEE, Mexico-Germany, 2020.

[46] Francisco Bulnes, "Controlabilidad y Señalización," Masterful KeyNote, UNITEC, CDMX, México, 1996.

[47] Stair Awards 2021, Science and Technology Academic, Innovation and Research Awards 2021.

[48] Francisco Bulnes, 2014, "Framework of Penrose Transforms on D_P-Modules to the Electromagnetic Carpet of the Space-Time from the Moduli Stacks Perspective." *Journal of Applied Mathematics and Physics*, 2, 150-162. DOI: 10.4236/jamp.2014.25019.

[49] Ghandhi, R.K.S., 1965, "Mao Tse-tung: His Military Writings and Philosophy". *Naval War College Review*. 17 (7): 1–27. ISSN 0028-1484.

[50] Dirlik, Arif, 2012, "Mao Zedong in Contemporary Chinese Official Discourse and History". *China Perspectives*. 2012 (2): 17–27. DOI:10.4000/chinaperspectives.5852.

[51] International Advisory Board in Research Foundation of India, n.d., Indiahttps://researchfoundationofindia.com.

[52] Bulnes, F., Fominko, S., "Local diffeomorphisms and smooth embeddings to gravitational field II: spherical symmetry and their breaking in the space-time," *Physics & Astronomy International Journal*, 2018, Vol. 2, (1), pp30-37.

[53] Francisco Bulnes, (Ed.), *Advances on Tensor Analysis and their Applications*, Intech, United Kingdom, 2020. ISBN: 978-1-83962-556-5, PRINT ISBN: 978-1-83962-555-8.

[54] Astronomical Observations by NASA, NASA, ESA, and the Hubble 20th Anniversary Team (STScI). https://noirlab.edu/public/es/images/STScI-gallery-1427a-2000x960/?nocache=true

[55] Astronomical Observations by NASA, ASA, ESA, H. Teplitz and M. Rafelski (IPAC/Caltech), A. Koekemoer (STScI), R. Windhorst (Arizona State University), and Z. Levay (STScI). https://science.nasa.gov/image-detail/hubble-ultra-deep-field/
[56] Francisco Bulnes, *Tratado de Matemáticas Superiores*: Análisis de Señales y Sistemas, ETC, Facultad de Ciencias, UNAM, 1998.
[57] Francisco Bulnes, "Doctoral course of mathematical electrodynamics," Internal. Proc. Appliedmath 2, SEPI-IPN, México, 2006, pp. 398-447.
[58] Heather Couper, Wikipedia, https://en.wikipedia.org/wiki/Heather_Couper
[59] Couper, Heather Anita. *Who's Who 2015*. Oxford University Press. Retrieved 2nd June 2015.
[60] Francisco Bulnes, Hall of Fame: Astronomy and Physics Award 2020, https://www.scholarshof.com/researcher/profile/?uid=389
[61] Francisco Bulnes, Y. Stropovsvky and V. Yermishkin, "Quasi-Relaxation Transforms, Meromorphic Curves and Hereditary Integrals of the Stress-Deformation Tensor to Metallic Specimens," *Modern Mechanical Engineering*, Vol. 2 No. 3, 2012, pp. 92-105. DOI: 10.4236/mme.2012.23012.
[62] Alfredo Sandoval-Villalbazo, n.d., *México Estancado en Porcentaje del PIB que Destina Investigación y Desarrollo*. http://ibero.mx/prensa/mexico-estancado-en-porcentaje-del-pib-que-destina-investigacion-y-desarrollo
[63] CEPAL's Economic Indicators and Projections, n.d., GDP growth forecast (%) according to CEPAL https://www.cepal.org/en/subtopics/economic-indicators-and-projections
[64] Post-Doctoral Project 2006-2007, 0001UCI/FBULNESUNAM, Nomination for the Budget for the year 2006-2007. First Visiting Professor at UCI.
[65] Francisco Bulnes, *Teoría de Algoritmos para la Maestría de Informática Aplicada*, Universidad de las Ciencias Informáticas (UCI), Habana, Cuba, Postdoctoral Project (Aplicaciones de la Teoría Infinita de Lie y sus Aplicaciones a Informática) 2006-2007.
[66] Francisco Bulnes, "Analysis of Prospective and Development of Effective Technologies through Integral Synergic Operators of the Mechanics," 14a. CCIA-CCIM, ISPJAE, Habana, Cuba, 2008.
[67] Francisco Bulnes, "Dinámica Lagrangiana y Calibres Electromagnéticos Aplicados a la Determinación de Curvatura de un Universo," 15 CCIA-CCIM, ISPJAE, Habana, Cuba, 2010.

[68] Francisco Bulnes, "Modelado Térmico de Transformadores Eléctricos a Través de su Evaluación por Operadores Integrales de la Mecánica II," 15 CCIA-SIDEGI. CUJAE, Cuba, Havana. 8th December 2010-December 10, 2010.
[69] UCI Archieve Pictures and Dr Bulnes's Archieve Pictures, visits: years, 2006, 2007.
[70] Ernesto "Che" Guevara, *El Diario del Che en Bolivia*, Editora Política/La Habana, Ciudad de la Habana, Cuba, 2000.
[71] UCI, *Notice on the intranet of the UCI and newspaper of UCI*, 2007. Archive Picture: Dr Francisco Bulnes giving a lecture on research.
[72] International Award on Educational Leadership to Dr Francisco Bulnes, from Liderazgo Hoy, A. C. https://iberomexico.com.mx/liderazgo-hoy/
[73] Sofía et al, *Forjando la Grandeza del Estado de México*, State of Mexico Goverment (Ed.) Vol. 2, Year, 2016.
[74] Francisco Bulnes-Aguirre, Eduardo Herna´ndez-Alvarez, M. C. Juan Carlos Maya-Castellanos, "Design of Measurement and Detection Devices of Curvature Through the Synergic Integral Operators of the Mechanics on Light Waves," Proc. ASME. IMECE2009, Volume 5: Electronics and Photonics, 91-102, November 13–19, 2009. https://doi.org/10.1115/IMECE2009-10038
[75] Francisco Bulnes, "The Anxiety Factor in the Dimension of the Symptom Space and their Elimination for Nanometric Actions of Quantum Mono-Pharmacists," Drug Discovery and Therapy World Congress 2013, 3-6 June, Boston Massachusetts, USA, 2013.
[76] Francisco Bulnes, E. Hernández, J. Maya, Volume 7: Fluid Flow, Heat Transfer and Thermal Systems, Parts A and B, ASME 2010 International Mechanical Engineering Congress and Exposition, November 12–18, 2010, Vancouver, British Columbia, Canada.
[77] Francisco Bulnes, F. H. Bulnes. "Quantum Medicine Actions: Programming Path Integrals on Integral Mono-Pharmacists for Strengthening and Arranging of the Mind on Body," Journal of Frontiers of Public Health 2012; Vol. (1), Issue 3, Hong, Kong, China, 2012 57-65. http://www.jphf.org/paperInfo.aspx?ID=21.
[78] Dr Francisco Bulnes, Editor-in-Chief, Journal on Photonics and Spintronics, Years, 2012-2020 http://researchpub.org/journal/jps/editorial%20board.html
[79] Francisco Bulnes (Editor-in-Chief.) Mini-Workshop on Nanoparticles, Photonics and Advanced Electronics, Journal on Photonics and Spintronics, Special Issue, 11th January 2018, Chalco, State of Mexico.

[80] Francisco Bulnes (Editor-in-Chief), *Journal on Photonics and Spintronics* ISSN: 2324-8572, Vol. 1, 2, 3, 4, 5 6 (special edition), Research Publisher, and USA Government, California, USA. 2012, 2013, 2014, 2015, 2016, 2017.

[81] Francisco Bulnes, *An electromagnetic ship designed by Lie-QED and their fermionic Fock space principles in superconducting to the future flights*, 4th International Conference and Exhibition on Mechanical & Aerospace Engineering, October 03-04, 2016 Orlando, USA.

[82] Francisco Bulnes, "Cohomology of Moduli Spaces in Differential Operators Classification to the Field Theory (II)," Proceedings of FSDONA-11 (Function Spaces, Differential Operators and Nonlinear Analysis, 2011), Tabarz Thur, Vol. 1, No. 12, 2011, 1-22.

[83] Dr Francisco Bulnes Archives and Comecyt Archives, Central Europe Archives. [83, 84, 85, 86]

[84] Dr Francisco Bulnes, *Baryogenesis until dark matter: H-particles proliferation*, Proceedings, 10th MATHEMATICAL PHYSICS MEETING: School and Conference on Modern Mathematical Physics: Belgrade, Serbia, September 9 - 14, 2019, 15-26 Contribution to 10th MATHEMATICAL PHYSICS MEETING, 15-26, 2020.

[85] Francisco Bulnes, n.d., "INTEGRAL GEOMETRY METHODS ON DERIVED CATEGORIES AND THEIR MODULI STACKS IN THE SPACE-TIME", Proceedings of 20th International Summer School on Global Analysis and its Applications: General Relativity 100 Years after Hilbert, Lepage Research Institute, Visegrad Fund, Slovakia, 20.

[86] Lepage Research Institute, 2023, Archives Program: http://www.lepageri.eu/summer-school/program.

[87] Doctorate Honoris Causa, Frankfurt, Germany, 2018, https://www.facebook.com/holaODAEE/photos/a.115854898441998/4670461836314592/?type=3

[88] ODAEE Archives, n.d., www.Odaee.com

[89] Francisco Bulnes, n.d., "La Verdadera Formación de la Educación: Personas hacia el Pensamiento Crítico", Aportes de la Paz, Interview from CEO: [https://www.youtube.com/watch?v=rshqGrdGQjQ].

[90] Verkelov, I., 2014, Pure and Applied Mathematics Journal. Special Issue: Integral Geometry Methods on Derived Categories in the Geometrical Langlands Program, 3, 12-19.

https://doi.org/10.11648/j.pamj.s.2014030602.13

[91] Francisco Bulnes & Fominko, S., 2021, Cohomology Groups, Currents and Dx-Schemes on ∂-Cohomology. Journal of Progressive Research in Mathematics, 18(3), 47-59. Retrieved from http://www.scitecresearch.com/journals/index.php/jprm/article/view/2082

[92] Francisco Bulnes., & Fominko, S., 2016, Dx-Schemes and Jets in Conformal Gravity Using Integral Transforms. International Journal of Mathematical Research, 5(2), 154–165. https://doi.org/10.18488/journal.24/2016.5.2/24.2.154.165

[93] Francisco Bulnes (Ed.) Recent Advances in Integral Equations, Intech, United Kingdom, 2019.

[94] Muskhelishvili, N. I. (Nikoloz I.) Radok, J.R.M. (trans), *Singular Integral Equations*, Noordhoff–Groningen Holland, 1953.

[95] A. Hammerstein, Nichtlineare Integralgleichungen nebst Anwendungen, Acta Math., 54, 1930, pp.117-176.

[96] Francisco Bulnes, Isaías Martínez, Rocío Cayetano, Andy Rodríguez, Isaí M. Martínez, Curvature as Obstacle to a Photo-Resistor Sensor of Illumining and Their Minimal Sensing Region Part II: Their Transitory Analysis and the Non-Dimensional λ, *International Journal of Sensors and Sensor Networks*. Volume 6, Issue 4, December 2018 , pp. 43-52. doi: 10.11648/j.ijssn.20180604.11

[97] Selinus University of Sciences and Literature, n.d., Doctorate Honoris Causa to Dr Francisco Bulnes from SUSL https://www.uniselinus.education/teachers/dr-francisco-bulnes-phd

[98] Amaqueme Regional del Estado de México, 2018, https://amaqueme.mx/2018/06/05/universidad-de-italia-otorga-doctorado-honoris-causa-a-docente-del-tes-chalco/

[99] Scientific Publications, n.d., World Journal of Electrical and Electronic Engineering, https://www.scipublications.com/journal/index.php/wjeee/editors

[100] Francisco Bulnes (Ed.), Post-Doctoral Seminar on Spectral Curvature and Spectrum of Curvature, Seminar Mexico-Ukraine-Russia, EAE, Germany-Spain, 2017.

[101] Francisco Bulnes and Olga Hachay (Ed.) *Advances in Complex Analysis and Applications*, Intech, United Kingdom, 2020. 978-1-83968-361-9, 978-1-83968-360-2 and 978-1-83968-362-6, 10.5772/intechopen.79799.

[102] Francisco Bulnes (Ed.) *Advanced Topics of Topology*, Intech, United Kingdom, 2022. 978-1-80355-094-7, 978-1-80355-093-0 and 978-1-80355-095-4.

[103] Journal of Nanotechnology in Diagnosis and Treatment. http://savvysciencepublisher.com/journal-of-nanotechnology-in-diagnosis-and-treatment/

[104] TESCHA Photograph Archive (Facebook): https://scontent.fmex32-1.fna.fbcdn.net/v/t1.18169-9/10580147_1445470072401276_8992340655925125016_n.jpg?_nc_cat=103&ccb=1-5&_nc_sid=cdbe9c&_nc_ohc=mC83iLBUKxoAX9zUebt&_nc_ht=scontent.fmex32-1.fna&oh=00_AT8ZrjE3USGn2YLpjHuyZdOky_41udLrkaC5NULDvaZX3g&oe=61E6F39C

[105] Dr Francisco Bulnes (Director, Chairman, Advisor, Ed.), Appliedmath I, II, III y IV, IPN-ESIME, IPN-ESFM, IMUNAM, FCUNAM, UAM, CINVESTAV, TecMonterrey, UAY, 2005-2008.

[106] Francisco Bulnes., 2012, Penrose Transform on D-Modules, Moduli Spaces and Field Theory. *Advances in Pure Mathematics*, 2, 379-390. http://dx.doi.org/10.4236/apm.2012.26057

[107] Francisco Bulnes, *Cohomology of Moduli Spaces on Sheaves Coherent to Conformal Class of the Space-Time*, Technical report for XLIII-National Congress of Mathematics of SMM, (RESEARCH) Tuxtla Gutiérrez, Chiapas, Mexico, 2010.

[108] Francisco Bulnes., 2014, Derived Categories in Langlands Geometrical Ramifications: Approaching by Penrose Transforms. *Advances in Pure Mathematics*, 4, 253-260. doi 10.4236/apm.2014.46034.

[109] Francisco Bulnes, 2013, Geometrical Langlands Ramifications and Differential Operators Classification by Coherent *D*-Modules in Field Theory. *Journal of Mathematics and System Sciences*, 3, 491-507.

[110] Francisco Bulnes, 2009, *On the Last Progress of Cohomological Induction in the Problem of Classification of Lie Groups Representations*. Proceeding of Masterful Conferences, International Conference of Infinite Dimensional Analysis and Topology (Ivano-Frankivsk, Ukraine), 1, 21-22.

[111] Knapp, A.W. and Wallach, N., 1976, Szego Kernels Associated with Discrete Series. Inventiones Mathematicae, 34, 163-200. http://dx.doi.org/10.1007/BF01403066.

[112] Francisco Bulnes, 2013, Orbital Integrals on Reductive Lie Groups and their Algebras. Intech, Rijeka. http://www.intechopen.com/books/orbital-integrals-on-reductive-lie-groups-and-their-algebras/orbital-integrals-on-reductive-lie-groups-and-their-algebrasB
[113] Francisco Bulnes, *Deep Study of the Universe through Torsion*, Cambridge Scholars Press, United Kingdom, 2022. ISBN: 1-5275-8728-2, ISBN13: 978-1-5275-8728-1.
[114] Dr Helen Thrift at Trinity College, 2019.
[115] Francisco Bulnes, 2020, Derived Tensor Products and Their Applications. In (Ed.), Advances on Tensor Analysis and their Applications. IntechOpen. https://doi.org/10.5772/intechopen.92869
[116] Adams J.F. "Stable homotopy and generalised homology." In: Chicago Lectures in Mathematics, MR 53 #6534. Chicago, IL: University of Chicago Press; 1974.
[117] Francisco Bulnes, 2021, "Motivic Hypercohomology Solutions in Field Theory and Applications in H-States," *Journal of Mathematics Research*; Vol. 13, No. 1, 31-40.
[118] Letters of engineering and postgraduate from physics and mathematics students (PhD Students and young PhD) from India, the Middle East and Africa to Dr Francisco Bulnes, 2013, 2014, 2016. of IINAMEI A. C. (Investigación Internacional Avanzada en Matemáticas e Ingeniería A. C) www.iinamei.com.mx.
[119] Letters of engineering and postgraduate in physics and mathematics students (PhD Students and young PhD) from India, the Middle East, and Africa to Francisco Bulnes, 2017-2021, 2022, of IINAMEI A. C. (Investigación Internacional Avanzada en Matemáticas e Ingeniería A. C) www.iinamei.com.mx.
[120] Ajdani, S. M., Francisco Bulnes., 2022, "Vertex Decomposability of Path Complexes and Stanley's Conjectures", in Francisco Bulnes (ed.), *Advanced Topics of Topology*, IntechOpen, London. 10.5772/intechopen.101083.
[121] Ajdani S., Stanley Cohen-Macaulay Modules, *Transnational Journal of Pure and Applied Mathematics*, (Dr Francisco Bulnes, Editor-in-Chief) Vol. 2, Issue 1, 2019, 37-47.
[122] Science Publishing Group, n.d., Integral Geometry Methods on Derived Categories in the Geometrical Langlands Program: Special Issue: *Pure and Applied Mathematics Journal*, https://www.sciencepublishinggroup.com/specialissue/specialissue info?specialissueid=141005&journalid=141

[123] Sarwat Aziz Abbasi, "Interview with Dr Francisco Bulnes," Bensci, Bentham Science, Benci, UAE, 2012.

[124] Francisco Bulnes, "Combination of Quantum Factors in Integral Monopharmacists and their Actions in Cellular Regeneration and Total Cure," ICDDT 2012, Dubai, United Arab Emirates.

[125] Francisco Bulnes. H. Bulnes-González, E. Álvarez, J. Maya, F. Monroy, "Integral Medicine: New Methods of Organ Regeneration by Cellular Encoding through Path Integrals Applied to the Quantum Medicine," Journal Nanotechnol. Eng. Med 1(3), 031009, 18th August 2010, (7 pages) doi:10.1115/1.4001879.

[126] Francisco Bulnes, H. Bulnes, E. Hernández, J. Maya. "Diagnosis and Spectral Encoding in Integral Medicine through Electronic Devices Designed and Developed by Path Integrals," Journal of Nanotechnology in Engineering and Medicine ASME, USA, 2011, pp 021009(2) 10.

[127] Francisco Bulnes, "The Anxiety Factor in the Dimension of the Symptom Space and their Elimination for Nanometric Actions of Quantum Mono-Pharmacists," Drug Discovery and Therapy World Congress 2013, 3-6 June, Boston Massachusetts, USA, 2013.

[128] Jung, Carl Gustav (2000 [2ª edición 2011]). Obra completa. Volumen 4. Freud y el psicoanálisis. 9. Ensayo de exposición de la teoría psicoanalítica (1913/1955). Traducción Ángel Reparaz. Madrid: Editorial Trotta. ISBN 978-84-8164-394-7/ ISBN 978-84-8164-395-4.

[129] Francisco Francisco Bulnes. H. Bulnes, D. Cote. "Symptom Quantum Theory: Loops and Nodes in Psychology and Nanometric Actions by Quantum Medicine on the Mind Mechanisms Programming Path Integrals," *Journal of Smart Nanosystems in Engineering and Medicine* 2012; 1(1) 97-121.

[130] Francisco Francisco Bulnes, H. Bulnes-González, 2015, "Quantum Developments in Nanomedicine: Nanocurative Actions by Soft Photons Sources and their Path Integrals," A. Seifalian, A. de Mel, D. M. Kalaskar (Eds.), Open Central Press, United Kingdom.

[131] Francisco Bulnes, *Mathematical Nanotechnology*, Generis Publishing, Slovenia, 2021. ISBN: 9-781639-02794-1

[132] Francisco Bulnes., 2013, Mathematical Nanotechnology: Quantum Field Intentionality. *Journal of Applied Mathematics and Physics*, 1, 25-44. doi: 10.4236/jamp.2013.15005.

[133] Francisco Bulnes. Martínez, I. Zamudio, O. Navarro, E., 2020, *Kinematic-Energy Measurements of the Torsion Tensor in Space-*

Time. In: Advances on Tensor Analysis and their Applications, 1st ed.; Francisco Bulnes., Ed., Intech, https://www.intechopen.com/online-first/kinematic-energy-measu Bulnes F, Editor. Intech: London, UK, 2020. DOI: 10.5772/intechopen.92815
[134] Robin Santos Doak, *Galileo: Astronomer and Physicist*, Capstone, 2005.
[135] The Holy Bible, (New International Version, Gen. 2.1).
[136] James Clerk Maxwell, 1873, *A Treatise on Electricity and Magnetism*, Vol. 1: 001, Vol. 2. 002, Dover, ISBN 10: 048646119X ISBN 13: 9780486461199
[137] Ravaioli, Fawwaz T. Ulaby, Eric Michielssen, Umberto, 2010, *Fundamentals of applied electromagnetics* (6th Ed.). Boston: Prentice Hall. p. 13. ISBN 978-0-13-213931-1.
[138] International Meeting of Algebraic Geometry and Mathematical Physics 2016 (AGMP2016), Tromso University Archives.
[139] Francisco Bulnes, "Derived Categories and their Deformed Versions through Integral Geometry Methods II", Algebraic Geometry and Mathematical Physics 2016 (AGMP2016), Research Report.
[140] Prof. Dr h.c. Gert-Martin Greuel, Director (2002-2013), Oberwolfach Institüte, Black Forest, Germany https://opc.mfo.de/person_detail?id=1428
[141] Dr Félix Recillas-Juárez, 80th Aniversary Fest, https://paginas.matem.unam.mx/matematicos/matematicos-r-z/matematicos-r/felix-recillas/802-una-conversacion-con-feliz-recillas-juarez-a-sus-80-anos
[142] Francisco Bulnes, H. Bulnes-González, 2019, "Quantum Developments in Nanomedicine: Nanocurative Actions by Soft Photons Sources and their Path Integrals," A. Seifalian, A. de Mel, D. M. Kalaskar (Eds.), Open Central Press, United Kingdom.
[143] Francisco Bulnes, V. Yermishkin and E. Toledano, "Constitutive Equations of the Stress-Strain Tensor for a Metal Speci- men Rehearsal in Quasi-Relaxation Regime and Their Generalised Functional of Energy," Proceedings of the 2nd CIMM, Vol. III, Department of Mechanical Engineering, Universidad Nacional de Colombia, Bogota, 8-9 October 2009, 12-0001-1-0010.
[144] Francisco Bulnes, J. Maya and I. Martínez, "Design and Development of Impeller Synergic Systems of Electromagnetic Type to Levitation/Suspension Flight of Symmetrical Bodies,"

Journal of Electromagnetic Analysis and Applications, Vol. 4 No. 1, 2012, pp. 42-52. doi: 10.4236/jemaa.2012.41006.

[145] Francisco Bulnes and A. Álvarez, "Homological Electromagnetism and Electromagnetic Demonstrations on the Existence of Superconducting Effects Necessaries to Magnetic Levitation/Suspension," *Journal of Electromagnetic Analysis and Applications*, Vol. 5 No. 6, 2013, pp. 255-263. DOI: 10.4236/jemaa.2013.56041.

[146] Francisco Bulnes, 2015, A Lie-QED-Algebra and their Fermionic Fock Space in the Superconducting Phenomena, Selected Topics in Applications of Quantum Mechanics, Prof. Mohammad Reza Pahlavani (Ed.), InTech, DOI: 10.5772/59078. Available from: https://www.intechopen.com/books/selected-topics-in-applications-of-quantum-mechanics/a-lie-qed-algebra-and-their-fermionic-fock-space-in-the-superconducting-phenomena

[147] Veinik, A. I., Thermodynamics of Real Process, Minsk, 1991. (In Russian.)

[148] Ramirez, M, Ramírez, L, Ramírez, O, Francisco Bulnes. "Field Ramifications: The Energy-Vacuum Interaction that Produces Movement," *Journal on Photonics and Spintronics*, Vol 2, no. 4, November 2013, 4-11.

[149] Francisco Bulnes, "Topological Quantum Diffeomorphisms in Field Theory and the Spectrum of the Space-Time, World Academy of Science, Engineering and Technology," *International Journal of Physical and Mathematical Sciences*, London, United Kingdom, Vol:11, No:5, 2017.

[150] Francisco Bulnes, S. M. Ajdani, Homotopies to Diffeomorphisms in Field Theory, *JP Journal of Mathematical Sciences*, Volume 30, Issues 1 & 2, 2021, Pages 1-18.

[151] Francisco Bulnes, 2013, Quantum Intentionality and Determination of Realities in the Space-Time Through Path Integrals and Their Integral Transforms, Advances in Quantum Mechanics, Prof. Paul Bracken (Ed.), InTech, DOI: 10.5772/53439. Available from: https://www.intechopen.com/books/advances-in-quantum-mechanics/quantum-intentionality-and-determination-of-realities-in-the-space-time-through-path-integrals-and-t.

[152] Alain Dister, *Le Rock Anglais*, Editions Albin Michel, Paris, France, 1973.

[153] Trips of Dr Bulnes as Mexican Scientist Delegate in many countries in the world. Archives of Comecyt (State of Mexico Consulate of Science and Technology).

[154] Francisco Bulnes, "Combination of Quantum Factors in Integral Mono-pharmacists and their Actions in Cellular Regeneration and Total Cure, Current Medicinal Chemistry International Conference of Drug Discovery and Therapy" 2012 Reports, Bentham Science, Dubai, UAE, 2012. https://docplayer.es/4966463-Informe-de-actividades-academicas-de-la-visita-a-emiratos-arabes-unidos-uae-united-arabic-emirates.html
[155] Francisco Bulnes, "Moduli Identities and Cycles Cohomologies by Integral Transforms in Derived Geometry," *Theoretical Mathematics and Applications*, vol.6, no.4, 2016, 1-12.
[156] Record Publons: Metrics and Statistics as Reviewer of Dr Francisco Bulnes, 2018.
[157] Francisco Bulnes, "Journal of Mathematics and Physics Reviewing Report I," *Science Domain Reviews Compilation*, United Kingdom, 2016.
[158] Michelle Musk, *Letters, emails and Interviews by Skype with Dr Francisco Bulnes*. Records and official letters (by post).
[159] Dr Helen Thrift, Communications and Archives of international visits, Wikimedia, Cambridge, United Kingdom 2016.
[160] Dr Francisco Bulnes, PhD, PostDocs, Doctor H. C., HonDSc, www.cestagi.com.
[161] Dr Francisco Bulnes, PhD, PostDocs, Doctor H. C., HonDSc, google-Académico.
[162] Francisco Bulnes, "Journal of Mathematics and Physics Reviewing Report II," *Science Domain Reviews Compilation*, United Kingdom, 2017.
[163] Award *Honorary Doctorate in Science* from IBC, Cambridge, United Kingdom, 6th March 2016.
[164] Helen Thrift, Research Department on Academic Relationship of Cambridge University, https://commons.wikimedia.org/wiki/Category:Mathematicians#/media/File:Mathematical_Research_Theory_of_Dr_Francisco_Bulnes,_by_Dr_Helen_Thrift.jpg
[165] https://www.youtube.com/watch?v=iZkRuEgBQfA
[166] Francisco Bulnes (Ed.) *Transnational Journal of Pure and Applied Mathematics*, Volume 1, Issue 1, 2019.
[167] Investigación Internacional Avanzada en Matemáticas e Ingeniería, 2019, Doctorate Honoris Causa from Claustro Doctoral Iberoamericano:

https://www.facebook.com/IINAMEI/videos/claustro-doctoral-iberoamericanodoctorado-honoris-causa-dr-francisco-bulnes-dire/361731167814224/

[168] Award *International accredited speaker* from CIC (Cámara Internacional de Conferencístas), USA. Award conferred in 2020.

[169] Reviewer Award for Journal of Electrical Science and Engineering ISSN: 2661-3247, para el 2022, https://tescha.edomex.gob.mx/sites/tescha.edomex.gob.mx/files/files/noticias%20investigacion/ESE%20-Reviewer%20Certificate%20Award.pdf

[170] Francisco Bulnes, 2016, *Electromagnetic waves in conformal actions of the group SU (2,2), on a dimensional flat model of space-time*. VI International Conference "Geometry, Dynamics, Integrable Systems" GDIS, Izhevsk, Russia.

[171] Law, S. N. Musk, M., 2018, *Great Men and Women in Science*, International Biographical Centre, Cambridge, Sara Rains (Ed.), UK. ISBN: 978-1-903986-44-8 (1st Edition).

[172] Helen Thrift, Michelle Musk, Short Reviews of the Francisco Bulnes's Mathematical Research Theory, *International Journal of Systems Science and Applied Mathematics*. Volume 4, Issue 1, March 2019, 13-17. doi: 10.11648/j.ijssam.20190401.12

[173] Rankings for Scientists, Dr Francisco Bulnes, World Scientist and University Rankings 2022, Francisco Bulnes AD Scientific Index 2022, Instituto Tecnológico de Chalco, USA, EU and Asia Rankings, 2022. https://www.adscientificindex.com/scientist/francisco-bulnes/804046

Further Reading

[174] Schwinger, J., "Particles, Sources, and Fields," Vol. (1), 3rd edition Advanced Book Program, Perseus Books, MA, USA, 1998.

[175] Verkelov, I. Francisco Bulnes. Goborov, R., "Fermionic Fock Space in the Superconducting Phenomena and their Applications," *Journal on Photonics and Spintronics*, Vol. 2, no. 4, pp19-29.

[176] A-Wollmann-Kleinert, Francisco Bulnes, "Leptons, the subtly Fermions and their Lagrangians for Spinor Fields: Their Integration in the Electromagnetic Strengthening," *Journal of Photonics and Spintronics* Vol. 2, no. 1, Feb. 2013, pp12-21.

[177] A-Wollmann-Kleinert, Francisco Bulnes, "Spinor Technology: the Field Formalism to the Duality between Quantum Particles and

their Waves," *Journal of Photonics and Spintronics* Vol.2, no. 2, May 2013, 9-18.
[178] Francisco Bulnes, "Integral Theory of a Universe," Internal, Proc. Appliedmath 2, SEPI-IPN, IM-UNAM, Mexico, 2006.
[179] A. Álvarez, Desigualdades de Hilbert e Integrales Orbitales de Flujo de Corrientes de Eddy para un Disco en Levitación, XLV Congreso de Matemáticas de la Sociedad Matemática Mexicana, 2013.
[180] Jornadas de Investigación Appliedmath 3 Dr Francisco Bulnes (Dir. Ed.), 2007.
[181] Nidal E. Hassan Taha, R. I. Nuruddeen, Abdelilah Kamal H. Sedeeg, Dualities between "Kamal & Mahgoub Integral Transforms" and "Some Famous Integral Transforms", *British Journal of Applied Science & Technology*, 20(3): 1-8, 2017.
[182] "Spare Parts", (2015) Drama Dir. S. McNamara, USA.
[183] "Stand and Deliver" (1989) Drama, Dir. Menéndez, USA.
[184] Garamszeghy, Mikos (1987). "Commodore in Europe: An International Comparison of Price and Availability" The Transactor. Transactor Publishing. 7 (6): 21–2.
[185] Simon and M. Reed, "Mathematical methods for physics," Vol. I (functional analysis)," Academic Press. NY., 1972.
[186] N. Joicey, *A Paperback Guide to Progress*: Penguin Books 1935–c. 1951. Twentieth Century British History 4#1 (1993): 25–56. doi:10.1093/tcbh/4.1.25.
[187] Adkin, Mark (2005). *The Trafalgar Companion: A Guide to History's Most Famous Sea Battle and the Life of Admiral Lord Nelson*. London: Aurum Press. ISBN 1-84513-018-9.
[188] Bodenheimer, R. (2011). "London in the Victorian Novel". In Manley, Lawrence. *The Cambridge Companion to the Literature of London*. Cambridge University Press. pp. 142–159. ISBN 978-0-521-72231-5.
[189] Francisco Bulnes, S. Humeini, "Topological Superconductors Characterising II: Curvature Energy in Hall-Spintronics Developments," *SciFed Journal of Spintronics and Quantum Electronics*, 1:1, 2018.
[190] Francisco Bulnes, A Tello, IX Congreso de la División de Dinámica de Fluidos, Instituto Mexicano del Petróleo, México, 2003.
[191] Francisco Bulnes, *Special Dissertations of Maxwell Equations*, unpublished book.
[192] O. Gonzalez-Gaxiola, Francisco Bulnes, "A generalisation of the Feynman-Kac formula through an operator algebra," *ADVANCED STUDIES IN THEORETICAL PHYSICS*, Vol. 7, 2013, no. 21-24.

[193] Francisco Bulnes (2015) "QED-Lie Algebra and their $\pounds -$Modules in Superconductivity," *Journal of Applied Mathematics and Physics*, 3, 417-427. doi: 10.4236/jamp.2015.34053.

[194] Francisco Bulnes, "Integral Transforms and Opers in the Geometrical Langlands Program," *Journal of Mathematics*, RP, Vol. 1, No. 1, USA, 2015.

[195] Feynman, R.P. (2006) [1985]. *QED: The Strange Theory of Light and Matter*. Princeton University Press.

[196] Francisco Bulnes, "Photonic Chains of Wave-links to the Quantum Communication," *Journal of Photonics and Spintronics*, Vol. 4, no. 3, pp10-14.

[197] C. Chevalley, *Theory of Lie groups*. Princeton University Press. 1946. 232.

[198] Francisco Bulnes, "Analytic Dissertations of the Sidereal Hyperspace" Masterful Conference given in UNITEC, Mexico-City, 1996.

[199] L. J. Mason, A Hamiltonian interpretation of Penrose´s quasi-local mass, Class. Quant. Grav. (6) L-7-L13, Cambridge, UK, 1989.

[200] L. J. Mason, J. Frauendiener, *Sparling 3-form, Ashtekar variables and quasi-local mass, Twistor, Twistor in Mathematics and Physics*, Cambridge, UK, 1990.

[201] G. F. Smoot, (2006). "Cosmic Microwave Background Radiation Anisotropies: Their Discovery and Utilisation". Nobel Lecture. Nobel Foundation. http://nobelprize.org/nobel_prizes/physics/laureates/2006/smoot-lecture.html. Retrieved 2008-12-22.

[202] 16. J. Krol, (2010b). (Quantum) gravity effects via exotic R^4, Ann. Phys. (Berlin) 19: No. 3–5, 355–358.

Figure 147. Dr Francisco Bulnes, with his great Cuban friend Dr Carlos Novo, in Havana, Cuba. Behind them appear the flags of Cuba and the University of Informatics Sciences. Dr Novo died in 2017.